Mining in the Old West

Sandor Demlinger

Schiffer Publishing Ltd

4880 Lower Valley Road, Atglen, PA 19310 USA

I dedicate this book to my wife Reva, my sons Keith and Glenn, and my daughter Allyson.

Library of Congress Control Number: 2005938931

Designed by Mark David Bowyer
Type set in Windsor BT / Cooper Lt BT

ISBN: 0-7643-2354-7
Printed in China

Published by Schiffer Publishing Ltd.
4880 Lower Valley Road
Atglen, PA 19310
Phone: (610) 593-1777; Fax: (610) 593-2002
E-mail: Info@schifferbooks.com

For the largest selection of fine reference books on this and related subjects, please visit our web site at
www.schifferbooks.com
We are always looking for people to write books on new and related subjects. If you have an idea for a book please contact us at the above address.

This book may be purchased from the publisher.
Include $3.95 for shipping.
Please try your bookstore first.
You may write for a free catalog.

In Europe, Schiffer books are distributed by
Bushwood Books
6 Marksbury Ave.
Kew Gardens
Surrey TW9 4JF England
Phone: 44 (0) 20 8392-8585; Fax: 44 (0) 20 8392-9876
E-mail: info@bushwoodbooks.co.uk
Website: www.bushwoodbooks.co.uk
Free postage in the U.K., Europe; air mail at cost.

Contents

Preface

I really began assembling this book twenty years ago. Since I am a photographer, I have been collecting photographs my entire life. One hobby of mine is model railroading, so collecting early photographs of mines, logging, old towns, and similar subjects came very naturally to me.

United States history is also a love of mine, as well as of my wife, Reva. We both enjoy collecting all kinds of historic photographs, and we started early enough that the price of the photos did not break the bank. We also found that photo postcards were a great source of this type of material. Many of these older cards can be one of a kind, almost like today's Polaroid photos.

Photographers of the old west and postcard photographers recorded many events, towns, streets, and industry. Most of the photographers took pictures as jobs came up, including family portraits. Burton Frashers, who worked out of Pomona, California, started by taking pictures of towns and every subject related to the west. His cards sold for pennies but he made a living. He even dared himself to drive through Death Valley in his old Buick with water bags draped over the engine on the hottest day of summer, and did just that. P.E. Larson, another early photographer, documented much history with his photos of Nevada's boom towns. Several photos by Frashers and Larson are featured in this book. There were countless other photographers who didn't leave their names on the pictures they took, so unfortunately I could not give them credit.

In this book of mining photographs you will find many images that have not been shown before. It also includes several important maps from our collection, documenting some of the unknown places where mining existed. These include a place in Death Valley called Skidoo, for example, which had its own newspaper at one time. You may get some eye-openers from the albums of photographs that people recorded to document the history of the mines.

We are certainly lucky that these images were not discarded in the trash like other pieces of photographic history that ended up unwanted. I think of glass plates from the U.S. Civil War as a sad example of lost history—many negatives ended up being used as hothouse windows for the flower trade. I also think of the niece whose grandfather photographed all of the great jazz masters in Chicago during the 1920s. She had to throw out boxes of glass negatives because there was no room to store them.

So we are grateful for these photos, which let us look back in time to see how mining influenced the growth of our country. Have a great time and enjoy this book.

—Sandor Demlinger

Introduction: Early Western Mining

A small piece of gold is so much heavier than the rocks and pebbles surrounding it—even though the latter may be larger in size—that a current of water washing the particles of dirt and rock away leaves the gold behind. Such was the case at Sutter's Mill in northern California, where the discovery of gold in 1848 set off a mining explosion that greatly hastened the development of the American West.

Upon hearing of this discovery, early miners flocked to those places where flakes of gold were concentrated and dug into the crevices, at first using small hand tools that enabled them to flick out the gold. Later they began using frying pans to wash away the dirt, leaving the gold particles behind. Soon larger pans with sloping sides came on the market, designed to assist the miners with panning. Another development was the rocker or sluice box. Its design resembled a wooden washboard that was placed at the bottom of a box. Water washed over the top of the box, leaving only the heavier gold behind. A similar box called the "long tom" was introduced by miners from the state of Georgia. The long tom was an inclined trough measuring twelve feet long, eight inches deep, and twelve to fifteen inches wide. It had an iron sheet with small holes at the lower end. This permitted the gold and finer sand to fall into a box below, while the larger rock could be cast aside.

By 1860, it had become harder to find places that harbored the gold, so the use of powerful hydraulics was initiated. Edward E. Mattison, who worked at American Hill, north of Nevada City in 1853, is credited with first making use of water under pressure to wash gravel into sluice boxes. That same year, a pioneer hydraulic mining system was used at Yankee Jims, located in Placer County, California, by Colonel W. McClure, who had visited Mattison and subsequently developed his own system.

The hydraulic mining system used canvas pipes attached to a nozzle resembling a garden hose. Great jets of water were directed against the tall banks of gravel and rocks along rivers that had gold in them. Thousands of cubic yards of gravel could be washed without hand labor using this method. However, the debris left behind became so objectionable that this method was stopped by an injunction in 1884.

Hydraulic mining was used in the region for twenty-five years. There were equipment changes as larger areas were discovered and stronger iron or steel pipes replaced the canvas hose. Large heavy metal monitors were used in place of smaller ones and the addition of bank blasting was a common sight. In this case, a T-shaped spike was driven into the banks at the top and black powder or dynamite was inserted into the T's ends. This instantly broke down the banks and it didn't take miners long to subsequently uncover the gold ore.

By 1887, pioneers were attempting to dredge gold from river beds—another method of obtaining the precious ore. A single lift elevator bucket was first used on the Yuba River in California. In 1898, W. P. Hammon began to dredge at Oroville. This dredge had open link buckets of three cubic feet and a tailing stacker. By 1901, electric power was introduced and steam used for the first time. This machine could dig sixty feet below water level. The most important California fields for dredging have been the Yuba River near Hammonton, the American River near Folsom, and the Feather River near Oroville.

Now that we have set the stage, let's go on to the "pot of gold" in the mining photographs that follow—many of which have never before been published. Each provides a fascinating record of a colorful and quite unique period in our history.

Chapter One

California

Among the first settlers to arrive in California by crossing the Sierras were members of the Bidwell Bartleson party, who entered near the Sonora Pass in 1841. As they had eaten the last of their oxen, they faced starvation when they reached the crest of the mountain range. Heading downhill was almost as bad as going uphill, with steep slopes, large rocks, lakes, and granite outcropping to contend with. They finally made camp at the middle fork of the Stanislaus River, not far from Knights Ferry.

This excerpt from a U.S. Army Topographical Engineers report, Sonoma, California, February 14, 1850, further describes some of the harsh conditions faced by early settlers to this region:

> Sir; Orders from headquarters of the Pacific division dated June 27, 1849 to take charge of an exploration from the upper Sacramento across the Sierra Nevada, to Humboldt river; main object of the expedition "being to discover a railroad route through that section of the country."
>
> With the killing of Captain W. H. Warner, by Indians, it now becomes my duty to command the topographical party and to submit to you the following report, which embodies as much information collected by Captain Norris as came to my knowledge. On the 24th of August, we left Deer Creek and entered the mountains, our party being about thirty-four strong, and having with us a wagon, cart, and eighty-six animals. We found the hills quite steep and the road very rough and rocky, so that with much difficulty we made nine miles, our mules being very poor, though the best we could obtain. The next day we again marched only nine miles and the mules with the cart giving out all together, it was dark by the time we made camp. We continued to measure the road with the odometer, but the slow-moving wagon and mules just slowed us with each day on the trail, and each hour more and more men became sick and soon more than half the party was unfit for duty. We met several men from the United States, and learned, to our surprise, that they were the advance of from ten to twenty-thousand emigrants, who were coming into the country on this trail. The advance parties were composed mostly of men who had left their trains, under a false impression as to the distance to the valley, and they were in a starving condition. They came to us in such numbers, begging for provisions, that our supply was materially diminished. As soon, however, as we met the trains, we found the emigrants generally pretty well provided. We succeeded in hiring several of them to replace those of our party too sick to advance; and leaving seven men in charge of the surgeons, and sending back our wagon and cart by one of the emigrants, with everything that we could dispense with, we left our camp on the fourth of September, and on the evening of the fifth we reached the headwaters of Deer creek, about fifty miles from Lassen's. Here we left behind five more men, too sick to proceed. It is a singular fact that, during the first ten days in the mountains, not a single one of our original party escaped an attack of fever…
>
> I have the honor, to present, with this report, a copy of the meteorological and astronomical observations taken upon the expedition, and a sketch of our route. A railroad route is possible through the area I have sketched from the mountains to the Humboldt river; there existed no serious impediment to the construction of a railroad. As the Lassen trail is now an emigrant trail from the United States, a list of distances from camp to camp, as far as we proceeded on it, may be useful, and therefore I submit it with the map.
>
> R.S. Williamson,
> Lieutenant United States Topographical Engineers,
> Bvt. Lieut. Colonel J. Hooker,
> Assistant Adjutant General, Pacific Division

"The delights of an overland Journey to California," a word to the wise for those anxious to partake in the developing "gold rush," was published in the *Norfolk Democrat*, Dedham, Massachusetts, on March 8, 1849:

> The following communication, which recently appeared in the *St. Louis Republican*, is from the hand of someone who has faced the elephant:

TO CALIFORNIA MEN

Permit me to caution your over zeal in going after gold. Take Warning before you engage in an enterprise beyond your age, strength, and means.

A long dreary road, of upwards of 2,000 miles lies before you, without houses, meat, or flour, and in many places without wood, water, and grass. You that start across the plains, by the time you reach Sante Fe, New Mexico, a distance of no less than 1,200 miles from St. Louis, will find your ambition and courage fail; and yet though Sante Fe is a paradise to what you will see afterward. By the time you are fifty days out, which is the time it will take for you get there, (or kill all your oxen and mules) you will wish yourselves back again at work in St. Louismark my words.

The following letter, from the pen of an officer in the U.S. Navy, gives a more particular account of the gold discovery and region:

Monterey, California Aug. 30th 1848

Gentlemen: The immense quantity of gold daily discovered in this region is almost incredible. It will be necessary to give a short history of the rise and progress of the Gold Diggings of Upper California, to make you understand something of the character and extent of the operations on foot for obtaining this "root of all evil." In February last, the renowned Sutter of Sutters Fort, had a number of men who were mostly Mormons employed in digging a race for a new mill. He was constructing a small one on the American Fork, a tributary, or, more properly speaking, a branch of the Sacramento river. On this stream, about thirty miles from the fort or "New Helvetia," there is a bend, across which, a distance of about two miles, the race was being dug.

During the progress of the work, but not until it was nearly completed, two boys who were engaged in the operation discovered some yellow particles in the sand, of so unusual an appearance that they were induced from curiosity to collect a quantity and take it home with them. Their parents submitted the yellow grains to the examination of Brannan, the Mormon leader, who unhesitatingly pronounced them gold, and immediately instituted a search himself, for more.

Colonel Mason has been up to the "Placer" with Lieutenant Sherman of the Topographical Corps and they have explored the whole region. He assures me that the reports in regard to the extent and quality of the ore are not in the slightest degree exaggerated, and that the gold is to be obtained there in almost unlimited abundance.

The officer goes on to talk further about the entire areawhich includes hundreds of milesand how rich it is in gold, then finishes with the following:

I have received orders this afternoon, to sail tomorrow for St. Barbary, St. Pedro and San Diego, and so must break off for the present and remain as ever, Yours Truly, S.E.W.

New York, Wednesday evening; Arrival of the Steamer Cherokee.

The steamer Cherokee has just arrived from Chagres, with $1,152,000 in gold dust and 142 passengers. She also brings California mails to the First of May.

The Tri weekly Placer Times has the following intelligence from the Sacramento diggings, which is the latest from that section of the gold region.

Mr. Johnbury has lately returned from the mines at Trinity. He went from here a short time ago, with a party of five men and seven mules;, the travel each way consumed sixteen days. He reports the gold as very abundant in the beds and bars of the streams, but in so very fine particles as to require the quicksilver machines, of which he took four back with him, to gather them.

Placerville does not flourish so well as it would if made the County seat of this establishment, instead of Coloma.

Prior to 1848, Calaveras County, California, was an unknown region, but by 1850 there were three counties in the area. With the 1848 discovery of gold at Coloma and other locations in the hills surrounding the town of Mokelumne Hill, a limit of only sixteen square feet per man was set; the ore was so rich in that area that everyone filled their sacks daily.

In August 1848, Jas. H. Carson, a Sergeant of Stevenson's Regiment, traveled south on a prospecting tour with man named Angel, which resulted in the settling of Angels Camp, where much gold was found. Soon more towns sprang up all over, with names like Railroad Flat, West Point, and Murphy's, to name a few. Eventually there would be more than 5,000 inhabitants in every town. There were of course some "bad apples," like the gang led by Billy Mulligan, who took control and forced many miners off their lands.

By 1898, 139 mines were located on top of French Hill, overlooking the town of Mokelumne Hill. This was one of the richest placer mines in the state, and was owned by a company of Frenchmen. Here, in part, is a letter written by August Sanno, describing what happened at his mine in 1851. He was one of the Frenchmen who were driven out by a mob.

> I was told to leave as I would be killed. I left with part of my gold and my partner had the balance. He was going to stay behind to see what was going on. I went ten miles to the Calaveras River, where I gave my $20.00 a day to help me recover my mine. I was riding a mule where that night the road was guarded by a lone sentinel.
>
> My mule was shot and I fell into a ravine by the road, and remained unconscious for a long time.
>
> My shoulder was broken and my knee was dislocated. When I came to, Mister King was near death and he died at Campo Seco.
>
> Crawling away from then all and night. A. Sanno.

Several of the early California mines were described as follows in Calaveras County's historical records of 1898:

> The South Paloma Mine has been sunk upon the south extension of the famous Gwin Mine. The Gwin Mine was one of the first quartz gold mines to be opened in California, with most of the richer gold coming from the free gold and the sulphuret which went to waste. This was later realized to have quantities of gold in the sludge; now sulphurets are all utilized.
>
> Angels Camp Mine was one of the richest and earliest settlements in the county, but the present prosperous condition of the vicinity is owing to the great work of the Hobart Estate owners of the Utica Mine, who has three claims in this area. The Madison who has a 40-stamp mill; the Gold Cliff which has 20 stamps; and the Utica Stickles where there are 120 stamps. Water is used for the power and the supply is owned by the company.

After the first wave of placer miners could no longer find the easy-to-pan gold flakes in the rivers around Sacramento, Columbia, and Placerville, they scattered up and down California, some to the desert, some south to San Diego County. By 1866, San Diego County had many small mines in operation, among them the Shenadoh, the Stonewall, the Pinacate, the Eureka, the Bladen, and the the Julian (the last still in operation, although used as a tourist attraction).

Moving further east toward Nevada and into the desert, placer miners founded many small towns by staking a claim for gold or silver in the area. One such discovery is told about the town of Bodie, California.

In 1859, a few prospectors and their burros were making their way over the high and low hills of the desert. They had traveled over the high mountains of the Sonora Pass, and coming down to a lower elevation they took a break. They each had a way of sniffing out any kind of ore and on this day William Bodey had a feeling. He took out his pick and began to dig. Looking closely at what he had dug up, the group saw to their amazement that it was gold flakes. Calling the area Taylor Gulch, they built a cabin near a spring. Panning gold seemed easy enough for these men, but winter was soon upon them and they hadn't realized how cold it would get. Only Bodey and Black Taylor were there when the storm hit. It was November and they were running short of supplies and would have to go by foot to Monoville for food and other provisions. On their return, they were overtaken by a blinding blizzard and the two lost their way. The supplies on their backs became too heavy as they fought to stay erect against the blustering winds and snow. Taylor realized that Bodey's strength was failing and tried to help him, but Bodey was too weak to go any faster and soon fell into the snow exhausted. Taylor tried to find him in the snow and called his name, finally stumbling onto his body. Lifting him up out of the waist-deep snow he staggered on for a while but had to drop Bodey to catch his own breath. When he tried to lift his friend again, he found it impossible. So he wrapped Bodey in a blanket and said he would come back for him. Somehow he reached their cabin, started a fire, made coffee, and ate some food. Putting on dry clothing he went out again to help his friend, calling his name over and over, but the wind and blowing snow were all he heard. It snowed for three days and when it finally cleared, he looked out into the sun-drenched snow banks but there was no spot that would mark the place where Bodey lay buried. Before spring arrived, Taylor searched the area and finally found the remains of his friend, his knife, and his pistol. He rolled everything in the blanket and buried it, placing a large rock on top for a headstone.

A problem with mining at this site was that very little water was available with which to wash the placer deposits. So in 1859, the first quartz ledge was located, to be named the Goodshaw. This location was above the place where Bodey had discovered the placer ground. On July 10, 1860, the miners of Bodey Diggings met and organized the Bodey Mining District.

By 1879, the population had grown to 12,000, including gamblers, businessmen, prostitutes, and of course miners. The Standard Mine was the largest in the district, although smaller finds were located on the hills.

Twenty miles from Bodie, up in the Sierra Mountains, was a mining town named Lundy. In 1878, W.J. Lundy began operating a sawmill on the lake named after him. He was supplying Bodie with lumber. Around the same time, three prospectors named W. D. Wasson, C.H. Nye, and L.L. Homer came upon some outcropping of ore and decided to have it analyzed. They found that there was gold in the mountains, so they organized the Homer Mining District of

Mono County. After a short time, many claims were made. Names such as Last Chance, Grand Prize, Savage, and many others were used, but the one that stuck was the May Lundy Mine. This was almost at the top of the mountain. Living and mining on such a steeply inclined mountain setting must have been very difficult, but people had to manage getting up or down if they planned to stay there. The small path to the mine seen in the photograph on page 43 shows that they did get there. The town itself had boarding houses, hotels, restaurants, and a Chinatown, and overlooked the Lundy Lake. The merchandise store was owned by man named Butterfield, The Lundy Hotel was run by a Mrs. Al Taylor, and the two large saloons were owned by George Fry and Jack Murray.

Another California mining area of note was the Rand Mining District, where gold was discovered in 1895 by three men: C.A. Burcham, F.M. Mooers, and John Singleton. The color of gold found at this site resembled that found in Africa, so the camp was named after a gold mining area in South Africa. The first milling of gold at the Yellow Aster Mine in this area was five and half tons of ore, and made $850.00. By 1898, a 30-stamp mill was producing three million dollars in gold. The area was first called the Rand Camp, but as soon as thirty people established the town it was named Randsburg. By 1900, the population had grown to 3,500 people and the town had schools, churches, and everything that made a community. During the winter of 1901, Randsburg experienced a smallpox epidemic with more than two hundred cases of the disease, but the Yellow Aster continued producing gold ore.

Not far from Randsburg was another mining site named Johannesburg, which was also settled and planned as a community in the Rand District. By 1896, it was an established town, having its own stores, hotels, homes, and a railroad from Kramer Junction, twenty-eight miles away. When the Yellow Aster began milling its own ore, it became independent of other mills in the surrounding area. Johannesburg established its own 10-stamp mill called the Red Dog Mill, and didn't need an outside source that required shipping out the ore.

"The Port of San Francisco June 1st 1849," by Geo. H. Baker Litho, from an original drawing by Geo.H. Baker and made expressly for *The New York Tribune*. About 200 vessels were detained at the port, their crews leaving for the mines. The view is from Rincon Hill, looking northwest and showing San Francisco Bay. The original lithograph is on heavy cream stock with black and pale green color, and is one of only three known copies.

Another mine in the Rand District was the Red Mountain Mine, established after a large silver strike in 1918. Two men, Hamp Williams, and Jack Nosser, were sent out by County Sheriff John Kelly to hunt for clay samples. They didn't find clay, but they did find a rich bed of silver. So while others were looking for gold, they stumbled into silver. By the time they went down ninety feet with a shaft they had taken more than $200,000.00 and the Kelly Mine was on its way. Other silver mines—such as the Silver King, Big Four Mine, and Silver Glance—soon followed, and all did very well. With miners moving in from all over, additional towns sprang up. One was called Osdick, named after a miner, but the post office got confused and decided to call the entire area Red Mountain. (It was officially named as such in 1929.) Red Mountain was soon turning into another mining frontier town with saloons and bars—the last in a long line of wild west towns.

Five miles from Randsburg a rich vein of tungsten was found in 1905, along the right of way of the Randsburg railroad. The town there was called Atolia. To mine twenty pounds of ore for $6.00 wasn't much, but with the railroad available, supplies and shipping were easy. With the advent of World War I, the demand for tungsten increased, since it was used to harden steel. As tungsten's price began to go up, so did Atolia's population—rising from 60 people to over 2,000. Atolia in 1916 was a good sized town with stores, saloons, and schools for the kids. But the war didn't last long and China began shipping tungsten to the U.S. cheaper then this country could mine it, so the boom days were over.

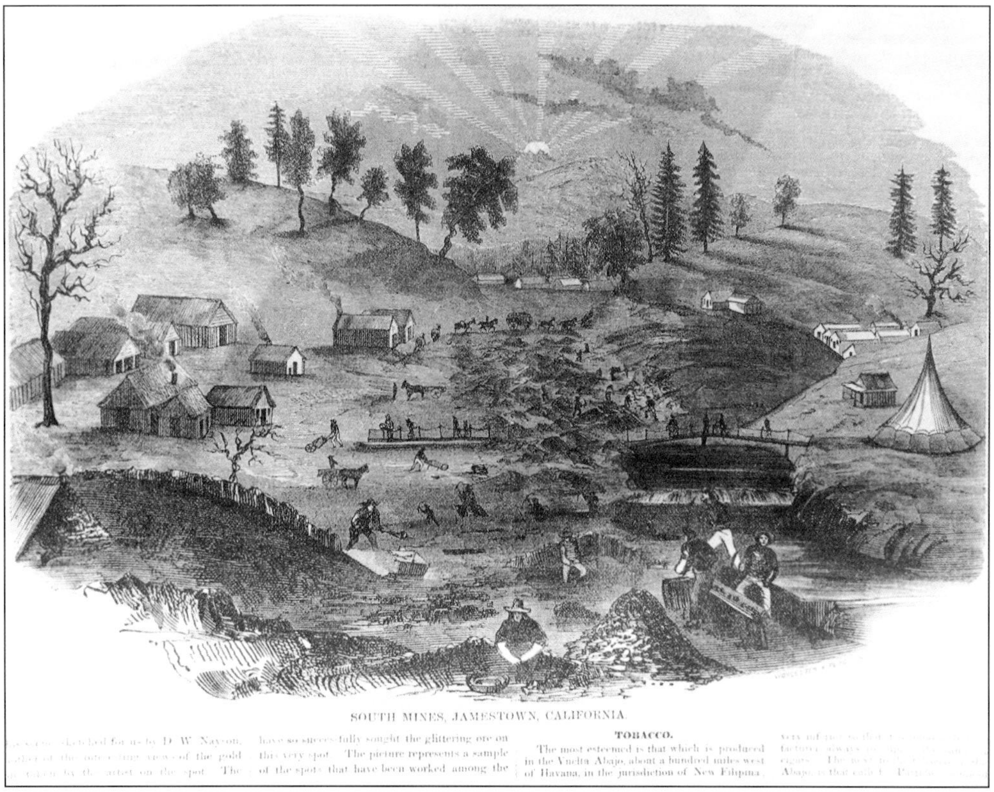

An early drawing from *Gleasons*, showing the South Mines in Jamestown, California.

An early view of the wooden buildings that made up the mining camp called Hangtown in 1849, and now named Placerville, California. In the early days, this area would have flooded or been buried in huge snowdrifts.

However there was still another boom going on, and that was for hard liquor. Since Prohibition was still in effect, people could make their own home brew and feel free to sell it. Nobody wanted to take on the rough and ready mining area, however, so anytime a raid was planned, the miners were always tipped off beforehand.

Interior of a mining shaft house, where a miner is being lifted out of the hole they had dug. The view is from an early *Harpers*.

An early stereo photo card showing placer miners, including Chinese, hard at work.

Placer gold miners panning for gold during the gold rush, with help from Chinese.
Courtesy of the California State Historical Society.

Using a string of Georgia long toms, these early miners have found their own place to work, away from the large crowds.

A early miner's sluice box with all the trimmings, including water pipes for washing and an angled drop sluice.

A copy from two tintypes showing the town of Volcano, California in the 1850s.

A fine example of placer miners that seems posed, but shows them throwing loose gravel into the sluice boxes and washing away all but the heavy gold pieces.

The band of miners from Eureka, California, in Calaveras County, looked like this during the celebration of their Jubilee.

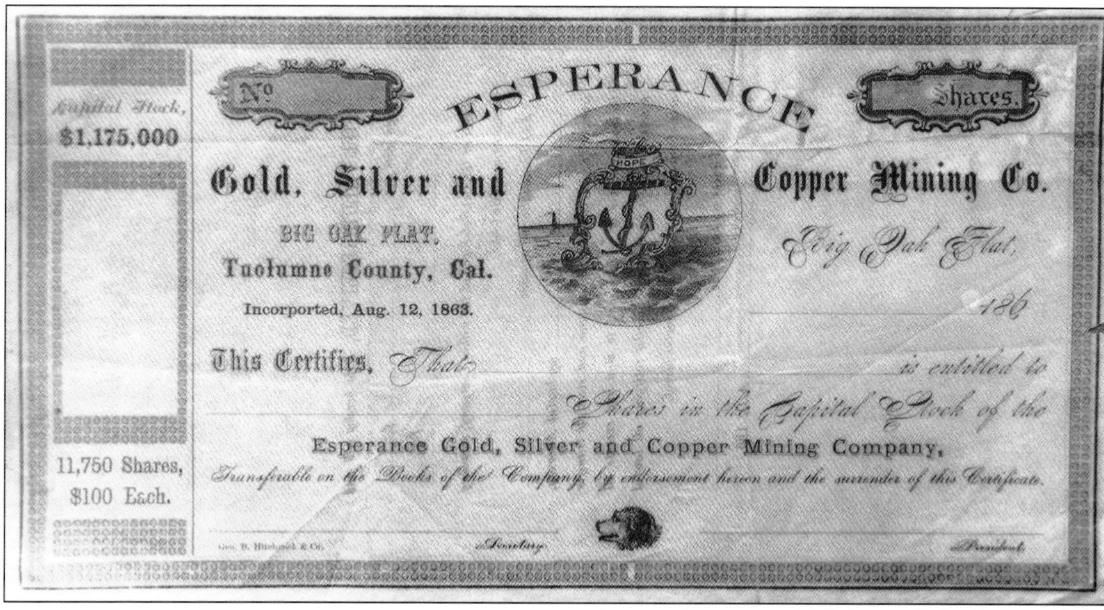

Capital Stock,
$1,175,000

No

ESPERANCE

Shares.

Gold, Silver and Copper Mining Co.

HOPE

BIG OAK FLAT,
Tuolumne County, Cal.
Incorporated, Aug. 12, 1863.

Big Oak Flat,
186

This Certifies, That is entitled to
Shares in the Capital Stock of the
Esperance Gold, Silver and Copper Mining Company,
Transferable on the Books of the Company, by endorsement hereon and the surrender of this Certificate.

11,750 Shares,
$100 Each.

Secretary.
President.

A example of stock issued by the Esperance Gold, Silver and Copper Mining Co. in Big Oak Flat, Tuolumne County, California, in 1863.

This old log cabin was built near a hydraulic mine site in the vicinity of Camp Baldy. Photo is credited to Alexander.

This old fellow is showing how he panned for gold during the gold rush.

These old timers sat for their portrait while prospecting for gold near San Andreas, California.

Titled "Old prospector on his way to the diggings." With a team of three mules and a wagon full of supplies, this man could last a few weeks in the mining camps around Chinese Camp.

A placer miner demonstrates the way he worked his pan looking for gold nuggets in a local riverbed.

Fully packed mules are about to set out for a mining expedition. Behind them stand some of the men who are backing this exploration.

In a place called "Happy Camp" we see the large pipe sections used with hydraulics. The water pressure was used to blast off huge chunks of earth and rock.

In this view of "Happy Camp," the miners—under the protection of a wooden roof—lift and feed piping to the mining site.

Three panoramic views of Truckee, California. The top photo shows the town's photographer setting up a shot of three mounted officers near his studio. In the middle photo, floats and cars lined up indicate that a parade is about to begin. The bottom photo shows homes and the Truckee Hospital (center of picture), with snow covering the street in front of them.

This very rare early photograph of Angels Camp Mine shows the small tracks and mine cars that are going in and out of the mine to the railroad dump site below. At the lower portion of the photograph, a small turntable to move the cars from one mine to another can be seen.

The main street of Nevada City, California in early 1900 looked like most early western towns. It included the important Assay Office and several bars.

The huge Argonaut Mine in Jackson, California, was billed as the deepest gold mine in the world.

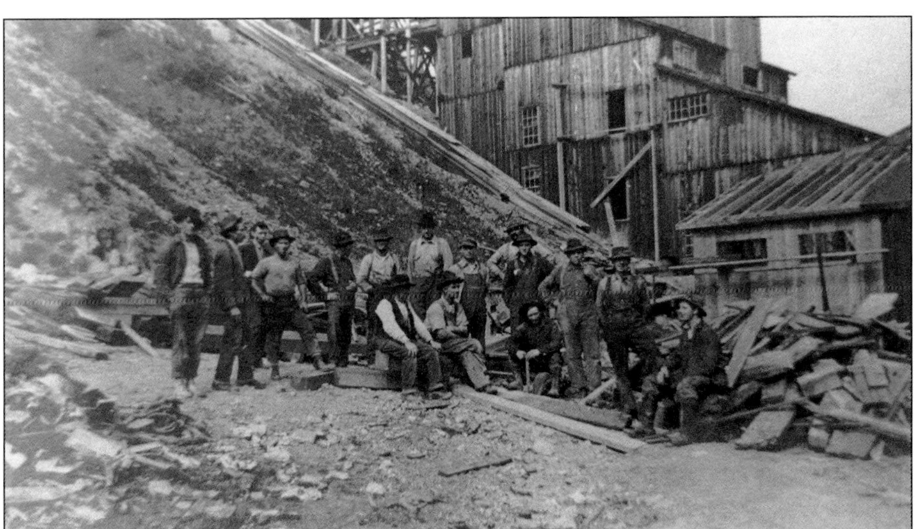

Lining up for their portrait, these miners get a break until the photo is taken. Behind them is the main structure at one of Grass Valley's many mines.

Portion of a map showing San Diego County's Mines and Minerals, dated 1886. Several mines are identified by small letters on the map: The letter A is for the Temecula; the letter B is for the site at Julian, now a tourist area; the letter C is for the Pinacate Mines; the letter D is for the Bladen Mines.

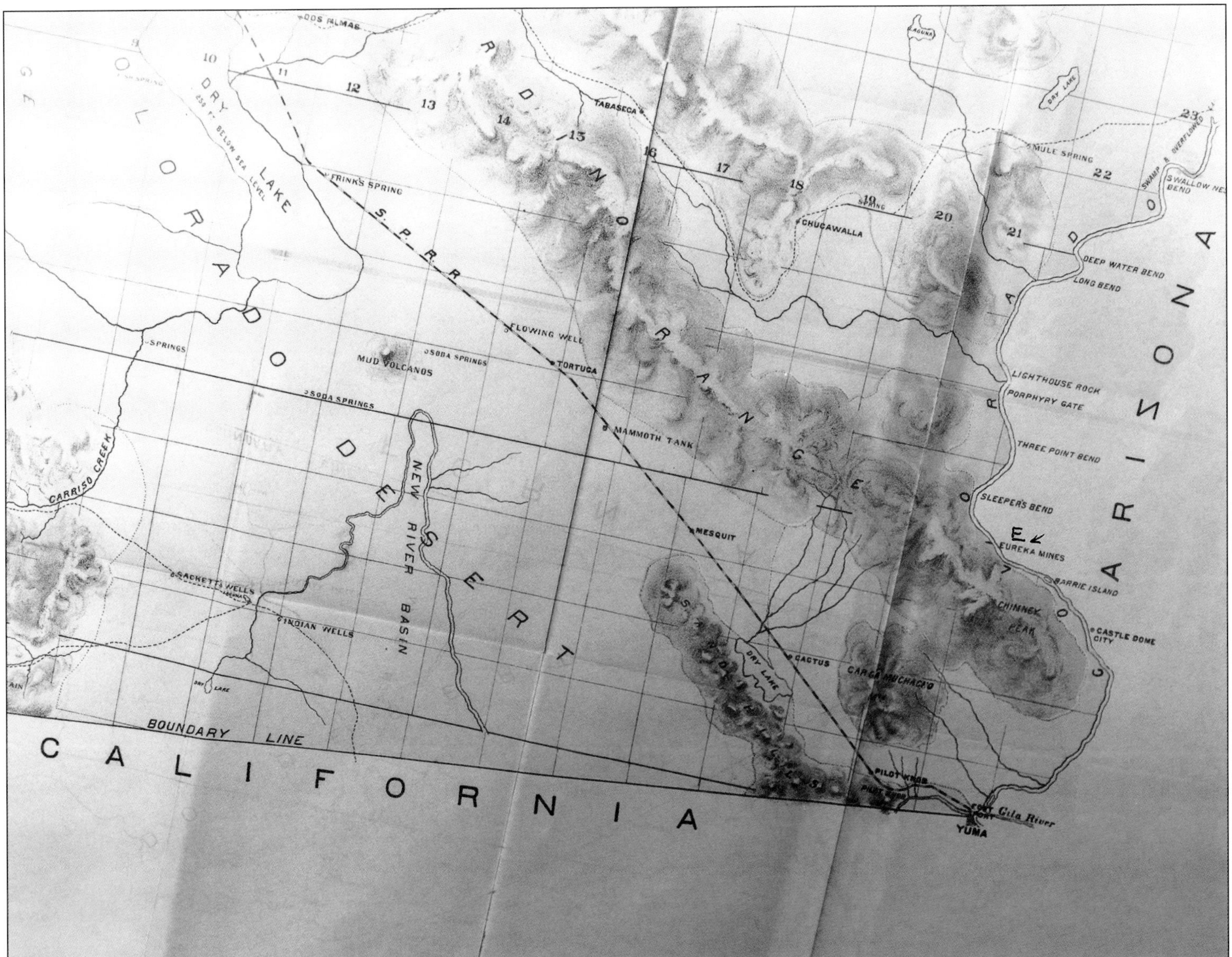

Another section of the San Diego Mines and Minerals map. On the right hand side is a small letter E for the Eureka Mines, near the Gila River.

The main street in the famous town of Julian, in San Diego County, where the mine still exists.

The Eureka Mine in San Diego County was located along the Gila River and this was one of the buildings found along the Gila River dock. These three photos were taken by a family member of someone who worked there.

This view was taken from the river, looking toward the Eureka Mines loading dock.

Another of the buildings located on the loading dock along the shore.

This photograph of the Banquet Mine, near the Mexican border in San Diego County, shows great details—including the day's wash hanging on a line.

One from a series of photographs taken by a miner's family member shows two miners taking a breather riding a set of mules. The mine site is in San Diego County, but remains nameless.

Laborers carrying supplies and dynamite up a steep hill in San Diego County.

A view from the top of a mine structure, showing the surrounding landscape and the working wheels atop a tipple.

A miner and his son, who already holds a lamp for the trip down below. The photograph is part of the series in San Diego County.

These men are erecting a mine structure, using pulleys to hoist up some heavy timber.

A small snapshot of a mine in San Diego County, showing part of the ore car trestles to the mine.

A rare look at the inside of a hut in San Diego County, where a chemist is testing ore from the mine.

The Black Canyon Gold Mine was located in Inyo County, California. This photograph shows the ore car tracks that led into the main tunnel, which went straight into the rocks.

Another view of the Black Canyon Gold Mine, showing other mine car tracks leading into the rock. This mine near Bishop, California was fed by the Carson and Colorado narrow gauge trains that went from Keeler to Laws in California, and Hawthorne to Mound House in Nevada.

Locomotive No. 6 of the Carson and Colorado was photographed at Keeler station in California. The year was 1907 and the crew was told to look away instead of frozen into the camera.

The town of Laws, early 1900s. Laws was a important stop for miners needing supplies before traveling into the desert or up and over the mountains.

The Laws railroad station for the narrow gauge trains of the Carson and Colorado railroad. The trains ran up until World War II, when the steel rails were ripped up and melted down for the war effort. The Laws depot still stands and is a historic sight, along with the Laws Museum, outside Bishop California. This photograph was taken by Forbes.

This early photograph of Bodie, California in 1905 shows the Standard Mine on the left and many of the houses and buildings. The writer of this card tell her auntie that the X (far right) is Mabel's house and the mine is where her father works.

This fine photograph by Forbes shows Bodie from a different angle. It includes the mine off in the distance, a main road in the foreground, and the Chinese and "red light" districts on the left.

A panoramic view of Bodie, long after the boom years. This photo was taken by Frashers about 1930.

A panoramic photo looking down one of the main streets in Bodie, showing the Miners Union building and other houses and structures.

This engine was named "Tybo" and was used on the Bodie & Benton Railroad. The locomotive was first shipped to Bishop, California, then dismantled and hauled by mule to Benton. It transported timber and supplies to the mines in the area. The photo was taken by Hawes, out of Carson City, Nevada.

During mining days, a crowd always gathered to witness two miners pounding out holes in a hard rock. Such was the case in Bodie, as more than half the town came to watch the two men in the lower right of the photo hammering away.

The best and most famous of bars was called the Bodie Bar. Here, two owners pose in front of their shiny bar.

A miner's home, with paper shades to keep out the hot rays of the sun, makes a fine portrait for Frashers's camera.

Stragglers walk up the steps of the Methodist Church on Sunday. Outside, a miner's fully packed mule waits patiently.

Aurora, California was just over the hill from Bodie and during its heyday even the likes of Mark Twain had a cabin there. This Forbes photograph shows the cramped little cabin he stayed in. The town was all but gone as soon as gold and silver prices hit a decline.

This panoramic photo by Burton Frashers shows the town of Aurora, California as it looked sometime in the 1930s. Frashers's son told me that he drove all over in the little Buick that appears (though it is hard to see) at the right hand side of this photo. Later, souvenir seekers took everything there was in Aurora, leaving only rotted wood planks.

A group of prospectors check their position before moving into the desert.

Published in the Wild Rose Mining District

THE SKIDOO NEWS

The Twentieth Century Gold Camp

VOL. 1, NO. 40 — SKIDOO INYO COUNTY, CALIFORNIA, FRIDAY, SEPTEMBER 20, 1907 — TEN CENTS

CAPITALISTS INVEST AT PANAMINT

Joseph Loughnane, the San Francisco manager and president of the Sonoma Express Co., L. R. Piper, a capitalist of Berkeley, and C. H. Brown, a mining expert of Ely, Nev., have just finished a three weeks' prospecting trip in the Panamints, where the latest discoveries have been made.

After a thorough investigation they purchased twelve claims and made several locations, showing their confidence in the camp. When interviewed by a reporter [illegible] expressing his opinion of the Panamint camp, said: "It is the best mineralized district that I know of in California or Nevada. It has an abundance of all kinds of ore, and what is needed mostly is a railroad and that is bound to come. We are preparing to commence work at once on our claims and expect to begin shipping ore within sixty days."

Messrs. Loughnane and Piper have left for San Francisco and will stop a day or two at Randsburg and Azalia to look over the mines at those places. Mr. Brown has gone to Goldfield and will return October first.

HART MINING CO. BUSY

A. T. Hall arrived from Rhyolite Tuesday with a complete camping and prospecting outfit. He is here in the interest of the Hart Mining Co. and will put in the season prospecting the claims belonging to that company and getting the property in shape for development. Mr. Hall has been mining in California and Nevada for over thirty years and has never before seen a camp with gold showings equal to Skidoo. He is confident of showing up something great on the Hart properties.

S. P. TO BUILD SOUTH FROM KEELER

The Motive

RICH LEDGE OF HOYT BROTHERS LEASE IS FOUND IN NEW SHAFT

Ore is of Higher Average Grade than in Older Workings. Strike Gives this Lease 210 feet of Stoping Ground, and again Proves Values at Depth.

Hoyt Brothers have struck it.

This constitutes the most important mining news of the week, and gives additional proof that Skidoo ores go down. And not only do they go down, but the average values at the present depth are considerably higher than in the more shallow workings of this property.

The Hoyt Brothers lease, on a block of the Skidoo Mines Co. ground, has for several months been looked upon as one of the big things of the camp. Bonanza ore was found near the surface and the ledge yielded very high grade average values down to a depth of sixty feet. The ledge having given every indication of permanence to that depth, prospect work was suspended and a permanent shaft, equipped with a gasoline hoist, was started which was expected to reach the ore on the dip of the vein at a much greater depth.

The expected result was realized this week. At the point where the ledge was cut by the new shaft it gives 210 feet of stoping ground on the dip of the vein. The ore here very closely resembles that of the Gold Bird found on the Granite Contact [illegible]

Big Advertisement for Skidoo

A rare newspaper dated September 20, 1907, from the desert town of Skidoo. The *Skidoo News* was the only paper for the short-lived mining town perched high above the Death Valley floor. Today, a marker that reads "Site of Skidoo" is all that remains of this tent city.

The famous Sutro Tunnel was begun in 1869, with the goal of digging a hole to the center of a gold-bearing mountain to save time. Sutro's eight-mile long railroad, pulled by a lone mule, actually worked.

An assayer checks the gold contents of samples sent to him by miners in Truckee and Placerville.

An early view of Calico, California from atop the mountain, showing the mine and surrounding town site. This photograph was made from a 35mm contact print and is the only such image known.

This street level photograph of Calico was taken in the early 1910s. It is later then the previous view, as the mine has been dismantled.

Another angle of the mining town of Calico, showing its one lone main street.

The Calico Saloon, in its early days. As with many other mining towns and camps, few women are shown in the photographs.

A final photograph of the character Deadwood Dick, who died May 5, 1930.

It looks like the entire town of Lundy has turned out for a group portrait. This scene shows the lake behind them, a few of the stores, and the Hotel Lakeview in 1900.

Miners throng to get good or bad news from the union building in Lundy.

Up in the Sierras, the May Lundy Mine looks like homes on a hillside overlooking tall, snow-capped mountains.

Another view of the May Lundy Mine, showing the narrow trail down or up from the mountain. Due to harsh cold winds, the mining site has been enclosed in a wood coverall.

The May Lundy Mine and town seen from the lake, which is near Monmouth, California today.

Barely visible is a high wire for the tram that went clear up the mountain to the mining site of the May Lundy.

A good example of the water chutes built to bring water to the mines, which were sometimes located in dry gulches. These were built high up in the mountains and some of the narrow structures traveled up to seventeen miles.

The area called the Rand District was made up of three principal mining towns: Randsburg, Johannesburg, and Red Mountain. This photograph shows the town of Johannesburg. The mine itself is way up in the background hills.

The town of Red Mountain, also on the Rand District map, had mining operations for silver. Along the main street are several bars and a view of the mine's headframe in the background.

This photo of Red Mountain was taken from the tungsten mining town of Atolia, California. When World War I ended, tungsten was no longer needed for steel, so the town itself dried up.

Another view of Red Mountain, showing the mine and other buildings, along with several tents used for supplies and housing.

The town of Randsburg, located in the high desert along the present day route of Highway 395. It was still mining gold as late as 1995. In this view, we see the pipe used to bring water to the mining site.

The Yellow Aster Mine in Randsburg looked like this while still in operation.

Tanks used in the mining operation at the Randsburg mine site.

View of a smaller operation in the Randsburg, California mining area.

In later years, dredges were brought in to salvage some gold from the waters at the Randsburg Mine.

This great photograph shows the 40-stamp mill of the Yellow Aster Mine in Randsburg, around 1898. The photo is from an original glass plate negative.

A rare photograph showing the tungsten mining town of Atolia.

In the middle of winter, Sol Ripinsky's general merchandise store (which appears to be located high in the Sierras) sold all sorts of animal furs.

Miners in the hills overlooking the desert break for lunch near their auto. In the second photo, one of the miners holds up a stick for sighting the location of a potential gold strike.

The workhouse and supply depot near the Fort Tejon mining site.

At the Fort Tejon mining site, just north of L.A., the mill is busy cutting timber for a water shed being built high in the hills. Also shown is a tram used to ship the cut lumber up and over the high ridge.

The local saloon offers relaxation to some busy miners near the town of Bakersfield, California.

This photograph shows the K Bros. store next to the Harrison Gulch Mine, which can be seen directly behind it.

This photo of a character named "Seldom Seen Slim" was taken in Ballarat, California, once a booming gold and silver mining center. Seldom Seen Slim, who reportedly settled in Ballarat in 1913 and stayed on long after the area became a ghost town, claimed to be the mayor, post master, dog catcher, tax collector, and traffic cop.

The Amador Hotel was a famous landmark in the mining town of Amador City, California. The hotel survived fires and mud slides during the early gold rush period.

Workers put the finishing touches on a very long waterway to be used for hydraulic mining in the area.

Two young boys sit on the waterway, which will wash away dirt, rock, and debris from the mining site.

A busy scene along the Sacramento River, showing mining camps and early mining equipment. A miner's early head frame can be seen on the far right and to the top of the photograph. Several miners are either washing or bathing in the river itself.

A photograph of Chinese Camp, a busy mining area in northern California. We can see that much of the timber has been cut and the wood used to shore up the tunnels in the mines.

Fully packed and ready to travel, these pack mules are about to leave for the desert. The prospectors' hopes for gold will go along with them.

The Empire Mine's headframe and machine shop in Grass Valley, California, serve as composition for the photographer.

Miners in Grass Valley pause for a moment to have their portrait taken.

After all the placer and deep mining had moved on to other areas, the mining companies called upon dredges to scoop out any gold or silver possibly still in the water.

Along the Folsom River, a dredge scoops the bottom.

A gold dredge hard at work along the banks of the Feather River, near Oroville, California. The photograph is from a early stereo card.

The Shasta dredge working along a narrow bend.

A dredge working along the Yuba basin, California.

A rear view of the Yuba basin dredge.

This photograph shows the main street of Jackson, California. In the background, we can just make out the headframe of the town's main mine.

An old time miner demonstrates the technique used in panning for gold. Taken at the famous ghost town that is part of Knotts Berry Farm, Buena Park, California, many years ago.

Chapter Two

Nevada

Of the many small towns that sprung up in the "mother lode" region, the town of Genoa, Nevada has an interesting history. In April 1849, Hampden S. Beatie and eighty men left Salt Lake City in covered wagons, arriving in the Carson Valley in June.

Beatie stayed on, building a log cabin on the present site of Genoa and establishing a trading post. Before the next winter, he returned to Salt Lake City. During his stay, however, Beatie made a few journeys over the Sierra Nevada Mountains into California for supplies. While he was away, his partner Abner Blackburn discovered gold in the small stream flowing from the south side of Sun Peak into Gold Canyon. There was little excitement about this find, as he and others felt no real amount of gold would be found at this site. Little did they realize that a few years later the area would become one of the greatest silver and gold bonanzas in the area.

Upon Blackburn's return to Utah, he told the Mormons about his find in the Carson Valley, so on their way back toward California, immigrants worked the gravels of Gold Canyon and the placer mining camp of Johnston grew up there.

In September 1850, the Utah Territory, which included parts of what would later become Nevada, was established. Mormons using the old Spanish trail through Las Vegas, Nevada into southern California discovered the Potosi Mine in 1855; the area is now called the Yellow Pine District of Clark County, Nevada. Placer miners from Johnston worked slowly up Gold Canyon and Six Mile Canyon and finally reached what would become known as the Comstock Lode in 1859. James Fennimore, known as "Old Virginia," named the site Virginia City, and together with several others who had worked the Gold Canyon, located the Gold Hill cropping.

In June of 1859, Peter O'Riley and Patrick McLaughlin, who also worked the Six Mile site, dug into the cropping of the Ophir bonanza and located it as a lode. The black mineral that clogged their sluices was found to be silver sulphide...the first silver mine in the United States had been discovered. The rush was on, and soon over 20,000 people came to Virginia City alone. Not all of them stayed, however; some moved on to discover more deposits in other areas of the state. The shift from mining in California to mining in Nevada came slowly, as miners were busy working the large mines in northern California or heading to British Columbia and as far away as Alaska.

The Central Pacific Railroad was extended from California to Nevada in December 1867, reaching Reno by May 1868. It joined the Union Pacific at Promontory, Utah, in May 1869, completing the first transcontinental railroad line. The Virginia and Truckee Railroad was completed from Carson City to Virginia City on November 12 of that same year, but did not connect to the Central Pacific at Reno until 1872. The Nevada-California Railroad was begun in 1880. In 1881, the Carson and Colorado Railroad was constructed from Moundhouse on the Virginia and Truckee Railroad to Hawthorne, and was extended later. The construction of these railroads was an important stimulus for the mining industry and its early prosperity.

This early period of prosperity lasted from 1869 to 1880. The period of decline began in 1881, as no important mines were being discovered in Nevada. In 1897, however, the Searchlight District was discovered and on May 19, 1900, Tonopah was discovered by James Butler. The Tonopah area had three mountains, Oddie, Butler, and Brougher, and ore was first found was at the foot of Oddie. Tonopah had a population of 6,500, and four banks. The wages paid to the miners were the highest in the country. Things went from very good to "what am I gonna do with all this ore." Shipment of top grade ore was always in demand, but the other stuff, which had too much of everything but gold in it, sat on the wayside, taking up space.

With Butler's discovery, more mining ensued, including the finding of gold by Harry Stimler and William Marsh, who founded Goldfield, Nevada in 1902. George Wingfield and Jack Davis together gave the Goldfield mining district a place in history. Wingfield was just a cowboy on the northern plains of Nevada until he met up with Senator George Nixon, who had a small banking business in Hawthorne. Since Wingfield was a dead eye shooter and a fair gambler, he staked the young man to some table stakes poker. George Wingfield, it is said, took all the money at this table and then some. Having heard about the rush of miners at the Goldfield site, he rushed in and bought considerable property, including land that fell on the main street. He even bought the Mohawk claim. Wingfield let George Nixon in on the Mohawk, and they became partners.

"Diamond Field" Jack Davis, of Idaho, Utah, Montana, and Colorado, was known as a gunman. He took out many of the thugs who tried running the miners' unions. At the Montezuma Club in Goldfield, Henry Weber, one of the new millionaires of the district, was told about Davis. A former governor of Idaho, a Mister Hunt, was asked if he knew him: "No I never met him, but I once issued a reprieve for him when he was about to be hanged. It is said that he and George Wingfield held off a gang of over one-hundred and fifty miners who came right to the door of their office in pursuit of a non-union miner."

By 1905, the railroad had reached Goldfield and the population had expanded to 8,000. Mills were built and a 100-stamp mill for the Goldfield Consolidated Mines was constructed. For many years, this mining operation was the largest gold producer in Nevada.

The Rawhide District lies to the South of Fairview and Wonder, just over the border in what is now Mineral County, Nevada. Charles Holman, a cowboy from Wyoming, named the site "Rawhide" on February 15, 1907. It turned out to be a phenomenal gold camp. By March 1, 1907, the first value of the ore was found. The claims were by C.C. Dunning and Jack Davis on what is known as Murray Hill. The first town site company was organized by Charles Walker and was called the "Rawhide Improvement Association." There were only twenty people living at the site at that time. The stampede to the camp didn't start until around October, because of rich strikes being made on another property called Balloon Hill. Capital began to flow in from all over after a thorough investigation by mining engineers, and Rawhide was on the map for a extended time.

Another important mining tale in Nevada history revolves around Rhyolite. And quite a tale it was—other mines have flourished and then died out, but the story of Rhyolite is like a sleight-of-hand trickster.

Most mining camps built housing with wood, tents, and even adobe, and their ruins are simply ghosts on the desert floor today. But Rhyolite was built to last—using brick and stone, huge buildings were being constructed. There was an opera house, a marvelous railroad station, churches, and a newspaper that printed

Many mines operated in the area around Virginia City, the "mother lode" region of mining in Nevada. This is a fine example of a newly built mine building. Photographed by J.H. Crockwell of Virginia City, Nevada.

the news about ...and this is where the story falls flat. No mine there had proven itself deeper then several feet. The town was built by those who didn't get in on the Goldfield strike and had something to prove.

This all started when Shorty Harris and Ed Cross struck genuine Bullfrog green ore a few miles southwest of Rhyolite. Its surface ores were richer then Goldfield's, or so thought many an investor, who gladly put money down on the table. However, a report by the U.S. Geological Survey, published in 1910, described the situation: "Whatever the expectations excited from time to time by the finding of superficial bunches of ore, there can be no doubt that the veins as a whole are to be classified as low grade. They are in no way comparable to the fame of Goldfield, Nevada." Situated on the very edge of Death Valley, the mines failed to produce, and the boom passed.

Yet some stayed on, with houses that had no furnishings, a restaurant that had no food, and an office building with three empty floors, intact. Had Shorty Harris come back to Rhyolite, he still would have been happy that he sold his share for drinks and a few hundred dollars.

Besides gold and silver, Nevada had mines producing copper, borax, and tungsten—all of which were sought after, even by the U.S. government, when the United States entered World War I. Some of these elements were necessary to produce higher strength steels.

A rare photograph taken inside a Virginia City mining building, showing some of the motors and pumps in operation. Photographed by J. H. Crockwell of Virginia City, Nevada.

Interior photograph showing the stamp mills and flat slabs that caught the crushed rocks.

Another interior view of mining operations, this one showing belt driven motors. The belts can be seen hanging from the ceiling.

An overall view of the Crown Point Mine on Gold Hill, in Virginia City, Nevada. This photo is from the Nevada State Museum's archive.

An early photo of Virginia City on a hot summer day. Two little girls pose for the photographer at far right.

Virginia City's B Street in 1915, as captured by the camera of Burton Frashers.

This picture was captioned "'C' street loves a parade." The original photograph was taken in Virginia City in 1903.

A fine photograph of mining entrepreneur James G. Fair, proclaimed Bonanza King of Virginia City, Nevada.

This photograph by Burton Frashers is captioned as the deserted main street of Aurora, Nevada. Yet other photos of the same place note that it is in California. This fine photo shows some of the still intact signs.

The early days of Goldfield, Nevada are illustrated here. In the background is the huge mountain that was explored and dug up for riches.

Another view of Goldfield's main street, showing more of the right hand side.

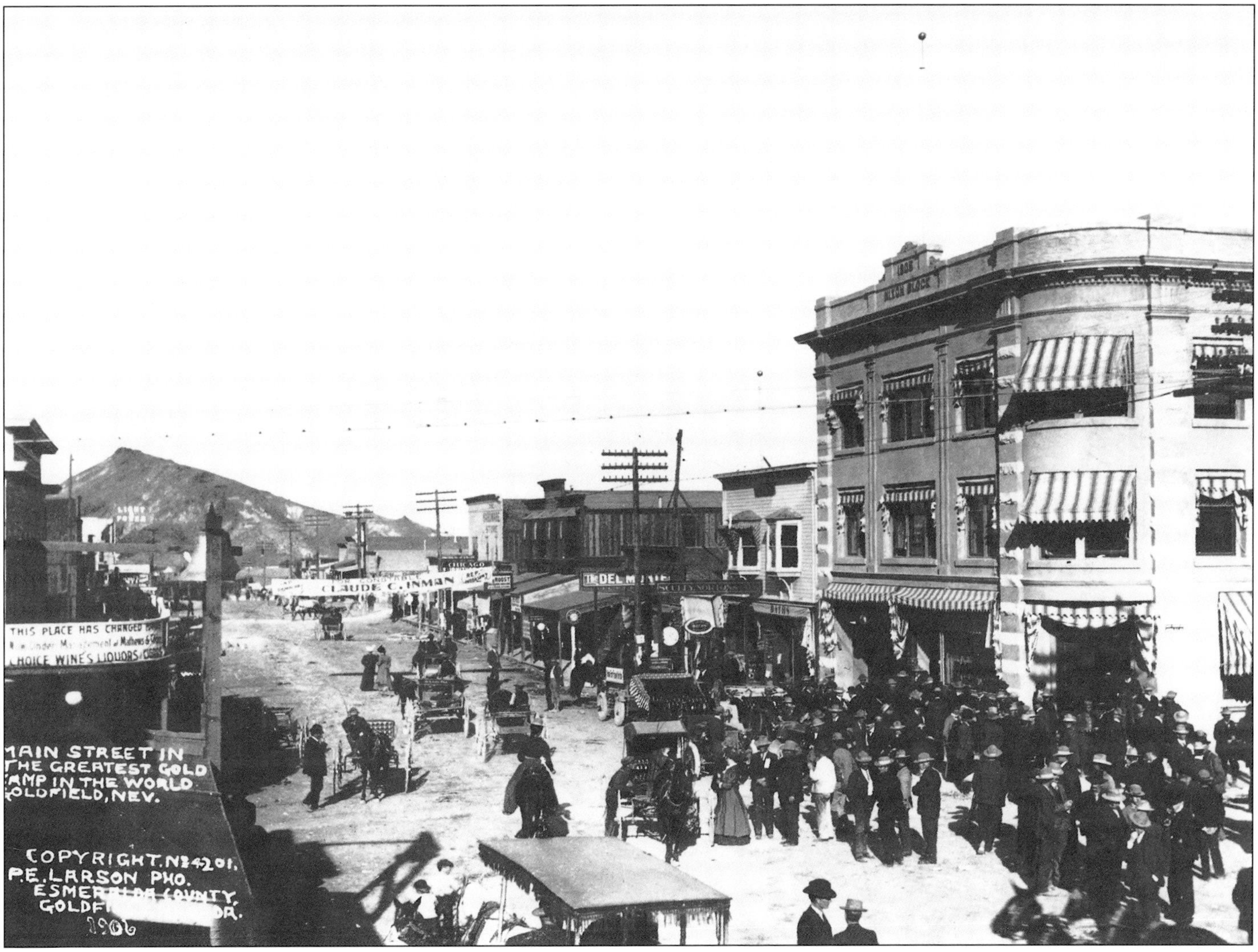

A busy street scene in Goldfield, taken in 1906 by P.E. Larson.

An armed guard stands by the State Bank and Trust Company's vault in Goldfield, Nevada. Looking on at left is the owner of the Mohawk Mine, D. Mackenzie. This photo is also by Larson, who photographed much of Goldfield's history.

The Mohawk Mine site, in Goldfield, Nevada. As we see, there are several headframes all working the area.

Mohawk Mines owner Mackenzie inspecting part of the mine, along with some of his foremen.

This fine view of mining operations at the Mohawk site shows the headframes and buildings.

The cold of winter has set in, but it didn't stop a photographer from taking this shot of Fifth Street in Goldfield.

Drilling contests were held in Goldfield, as in Bodie, California (see page 35). Here, 20,000 people have come to watch two men pounding a hole in rock. The only difference was, these guys meant business—judging by the large sledge hammer one of the men has raised up.

A teamster holds up a huge team of mules to be photographed on the main street of Goldfield, prior to their hauling boilers and other equipment on to the mines.

One of the principal owners of a Goldfield mine holds a large gold piece (front center), while his partners and their families sit for a photographic portrait.

Some important folks surround the talented beauty Lillian Russell, who sits while James Braddock the boxer looms high over her head. Braddock was later to box in Goldfield's outside ring.

Supply teams march into Goldfield.

As noted in this photo's handwritten caption, a monopoly caused the citizens of Goldfield to freeze in the cold as they waited to apply for coal and fuel. The price was $95.00 a ton, cash, for the "cheapest fuel in the U.S."

This large group of mules is leaving Goldfield with some heavy parts and equipment. In the background, we see the town and famous mountain.

Brave souls stay on in the winter at the ore sites of Rochester, Nevada, near Wimmemucca. Here the post office building provides some shelter, while mules wait outside.

A calm scene on Bridge Street, in Winnemucca, Nevada. This photograph may have been taken before any gold or silver was discovered in the area.

A busy street scene, with unloaded wagons and mule teams coming and going in the town of Winnemucca.

Miners working the hillsides of East Rochester, Nevada. Their tents give us a clue that cold weather is on the way.

Here is Gold Hill, Nevada and its working mines. The town itself is below the mound of dirt in the foreground.

This photo shows the small Bullfrog Gold Bar Mine, high in the Nevada hills.

Another small mine in the Bullfrog District of Nevada. The miner stands alone on the far left, posing with bags of treasure surrounding his camp.

One of the larger mines in the Bullfrog District was this Montgomery Shoshone Mine operation. The town nearest it was Rhyolite, Nevada and this photo was taken in March of 1907. The writer of this postcard asks, "Lovely town here, yes, no."

In Rawhide, Nevada, a postcard photographer took this fine example of a mine structure, including chemical tanks below the building and a long tramway at upper left.

Photographer Ned Johnson took this and other pictures of the growing town of Rawhide, Nevada in 1908. It began as a tiny tent town, but grew to one with large wooden structures, mines, and hundreds of people coming to the new boom town area.

Some of the machinery under the mine structure was made of wood, as illustrated here, while other key parts had to be constructed of steel.

Two images from photographer Ned Johnson, who photographed much of Rawhide's history. One shows the first saloon in Rawhide, built out of canvas to form a tent cover. The second is a very early shot of South Rawhide.

The Rawhide Queen Mines, several city blocks long and wide.

The arrival of cowboy Tex Rickard's huge drill for mining and water was cause for celebration in Rawhide. There were many stores opened for business on this day.

In 1907, Rawhide had few wooden buildings and consisted mostly of tents scattered about the hillside.

RAWHIDE OWL ONE.
Located on Hooligan Hill. Great Placer Values Are Found in Flat Shown in Distance.

A photo of the Rawhide Owl One mine site, located on Hooligan Hill. The caption tells us that there was great placer mining on the flat shown in the distance.

Nevada Street, one of the main streets in Rawhide, featured the Montana Café and a telegraph pole to facilitate important transmissions from San Francisco and elsewhere.

Steam powered this huge rumbling machine as it hauled large pieces of borax from the desert mills. The road, named Mule Canyon, was originally used by the twenty mule teams of earlier years.

Newer homes are seen in this late photograph of Silver City, Nevada, with some of the mines still visible in the background.

Not seen in the previous view is this abandoned mine site in Silver City.

These gents traveling in an old auto include two important people who helped build Nevada's mining industry. In the rear sits Jim Butler, who discovered Tonapah, while driving the car is Con Ahern, who owned the Crystal Bar in Virginia City. The third man could be their bodyguard.

It looks like things were buzzing in this view of Tonopah, Nevada, taken from a photo postcard. One of the mines at the site can be seen far in the background.

Another view of Tonopah shows several mining sites plus more of the housing in town.

Bragging that "Tonopah is the Silver Mining Center of Nevada and One of the Greatest Mining Districts in the Nation," this photo postcard served as a advertisement for the town.

This small jail house was used to hold criminals in Tonopah.

Displaying solid silver bars melted down from the ore in Tonopah, these miners and owners sit proudly by their treasures.

Pulling for all they are worth, these eighteen mules make their way with heavy machinery toward Tonopah.

On July 4, 1903, Tonopah, Nevada held a drilling contest. One contestant (bottom left corner of the picture) is about to swing down on the drill bit. Even a photographer on the roof of the Butler building has gotten into the excitement.

Two photographs were pieced together to create this panoramic view showing the remaining site of Rhyolite, Nevada.

A closer look at the few brick walls left standing at the site of Rhyolite, Nevada's once busy mining town.

The chemical tank at the McGill Mine in Nevada is smoking in the cold of winter.

The mining town of Ely, Nevada at its busiest time, with large brick buildings containing stores and shops.

This horse and mule team is having a hard time of it in the middle of winter. It was headed to a mine site in Nevada carrying very heavy equipment.

Young children face the photographer in this photo of a mining site somewhere in Nevada. The huge wheel for generating power from the rushing river can be seen in the background.

Miners stand outside a small mining operations building at Wadsworth, Nevada.

The writer of this photo postcard tells her friend that everyone has colds, and offers to send some peanuts. Eureka, Nevada, looked like this in early 1930.

Lovelock, Nevada was a small town but seems to have a schoolhouse and a bell in the steeple. Such bells in almost all mining towns served as warnings for fire or storms.

At Ruth, Nevada, the shops of the Nevada railroad were photographed. Signs identify each building's purpose, from the round house to the sand house. In the distance, we can make out the headframe of a mine.

Some of the heavy machinery used in the mining industry can be seen in this photograph. Large turbines were needed to run the stamp mills and for hoisting.

In 1913, a prospector stops to gaze at some rock formations in the desert near Schurz, Nevada.

While bringing out a wheelbarrow full of rock from within the site, a miner stops to have his portrait taken. The location is the Copperhill, Nevada mine area in 1919.

Chapter Three

Colorado

While more and more stakes were being discovered throughout Nevada and California during this early period, other western states attracted miners seeking their own rewards. They searched high and low over badlands and Indian battlegrounds looking for that fabled pot of gold. Sometime around 1872, J.B. Stansell and Abram Bergh put up the first house in Park County, Colorado, in a town called Alma. The town was established after the discovery of placer gold and was soon booming in one year, with a population of over 500 people all working hard for the dollar bill. All around Alma, some kind of metal was found: gold, silver, and copper ores were plentiful. By 1879, silver discoveries had also been made on the slopes of Mount Lincoln. Mount Bross, which is 14,000 feet high, also had miners climbing it to dig out the metal. Colorado was now the miners' boom state, and they flocked there by the thousands.

In 1879, Como, Colorado, was just an end-of-track town that had become a city of tents. It existed because railroad crews working there were busy laying tracks for the Denver, South Park, and Pacific Railroad. At this time, the mad mining rush was at Leadville, Colorado, and Como was a stagecoach stop on the way to that town. Traffic moving through Como also included freight wagons, mules loaded so much that their legs wobbled, and of course people with packs on their backs. With all this heavy traffic, Como became a town of over 6,000, and more were coming from a wasted gold town called Hamilton.

The original Como was actually located in a meadow at the site of King Park. When the railroad built a station three miles away in 1879, people moved there, and that new location became the present Como. For many years, starting in 1879, the South Park Coal Company (which also operated the King Mines) furnished the railroad with a steady supply of fuel. Late in 1879, Chinese miners were brought in from Fairplay, and this move brought about a small community war as the Italians who worked the mines ran the newcomers out of town. Como was a town without electricity and people used gasoline lamps to light the streets as late as 1963.

Speaking of Fairplay, the town's name has an interesting history. In 1859, when miners rushed to Pikes Peak and found that the best claims had all been taken, they headed into the mountains and some found good placer in Terryall Creek, just north of Como. But there again, the first arrivals got the best placers and the term "Turn Backs" was used for those who would go elsewhere. They moved across South Park to start a town of their own, and named it Fairplay Camp. Fairplay claims the oldest operating courthouse in Colorado. Its brick and stone walls served as shelter from the 1879 "fake" Indian uprising, started because a drunk shot his hat full of holes, then told all who would listen that the "Injuns were coming."

The Cripple Creek District of Teller County, Colorado is located on a hill and mountain range that ranges from 9,000 to 11,000 feet in elevation. The towns of Cripple Creek and Victor are located among these mountains. In 1891, gold was discovered in this area, which was first used by cattlemen as a range for their stock. There had been a small strike in 1874, when a little float of ore was discovered around Mount Pisgah, but nothing much came of that. In fact, the area looked nothing like a place where gold would be found. The hills were round and smooth, and covered with grass. There were no deep ravines cut into rocks, and the forest surrounding the area, although not large, hid the bedrock. In fact, the gold taken from the region was a dark brown color, not the usual golden color most miners would be used to seeing.

One of the first gold seekers in the Cripple Creek region was Robert Womack, who had worked the gold mines around Gilpin County with his father as a young boy. The Womack family acquired the land on which the Cripple Creek mines would later be located, but used it for cattle grazing. His father eventually sold the land, but Robert got a job with the real estate men who put young George Carr in charge and they both worked as range riders. Robert still had the feeling that somewhere on this land there was a lot of gold. He did uncover several veins of rich ore, and taking some samples with him, he turned up in Colorado Springs where he found that no one was very interested. Between 1880 and 1890, Womack—still working as a range rider—would uncover more gold samples and continue prospecting.

In December of 1890, two men, E.M. De La Vergne and F.F. Frisbee, came up from Colorado Springs to do some prospecting in the area. They found some good samples of ore, even though most of the countryside was deep in snow. Returning in February 1891, they found Womack at work at his claim in Poverty Gulch. Womack had not recorded his claim, but as soon as he saw these two men he located it, naming it "El Paso." Later, however, Womack was induced to give up his bond and option on the El Paso for $5,000.00. Thus, the man most responsible for the discovery of Cripple Creek got little out of his major gold strike.

Another man who stands out in Cripple Creeks history is Winfield Scott Stratton, who would become the wealthiest mining man in Colorado. Stratton was not originally a miner, he was a house builder. However, he did prospect for over twenty years and was familiar with mineralogy and geology. In May of 1891, Stratton, through conversations with Frisbee and De La Vergne, became interested in the Cripple Creek District. Camping there, he found specimens of very rich ore—but the rich lode still remained to be discovered. Near a spring at Willow Creek he found more gold specimens, so Stratton began digging trenches.

On the following Fourth of July, Stratton came to the realization that the outcrop was the lode. He returned to Cripple Creek and, locating the Independence and Washington claims, broke off some sample ores that he gave to his friend Troutman. The next day, Troutman gave him an assay certificate showing that this ore would run about $350.00 a ton. This was a fortune waiting to happen. Stratton sold the Independence for ten million dollars. This discovery was the beginning of the Cripple Creek mining district.

Labor problems and the number of new mines constantly opening led to a decline in production beginning in 1903. The last rich ore was found at the Cresson Mine in 1915. By 1925, visitors would travel to the area to see what a thriving, but abandoned, gold mining district looked like.

PROSPECTING ON THE UTE RESERVATION—AN OMINOUS MEETING.—DRAWN BY W. A. ROGERS.

Illustration from an early *Harpers* magazine, showing the plight of a prospector confronted by a Ute tribesman in Colorado.

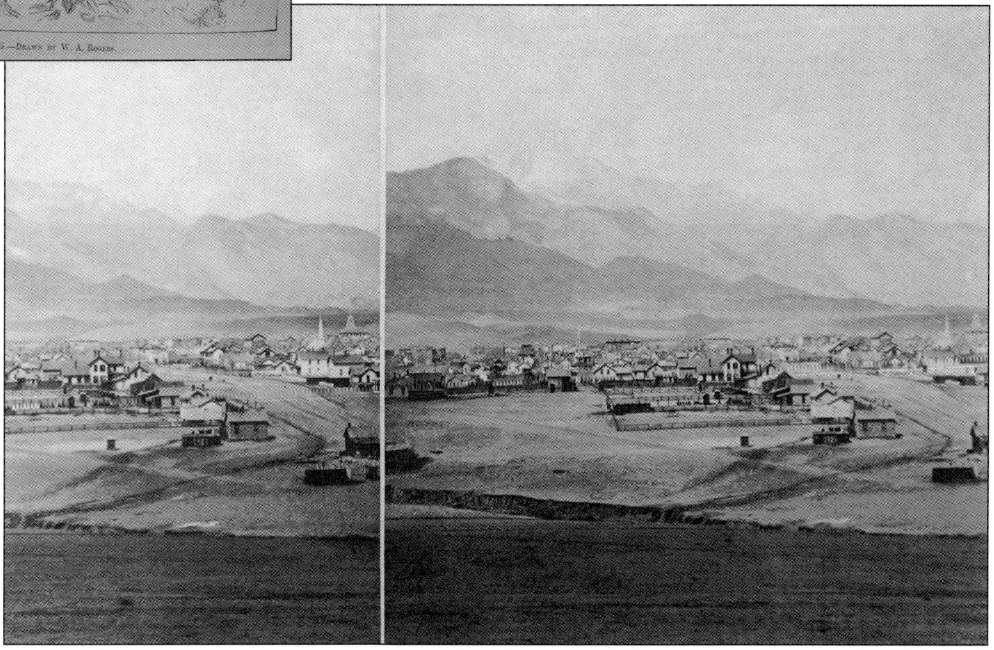

A stereo photo of very early Colorado Springs, Colorado.

This photograph, taken from a large stereo card, shows a mine site and several buildings somewhere in the Colorado mountains. The miners themselves stand on the far right.

Taken at the same mining site in Colorado, this photo shows construction of a processing plant. Over eighteen men are seen along with one lone women, who sits on her horse in the center foreground.

Colorado photographers used a lot of stereo cameras on location, since they were small and light weight compared to the larger glass plate cameras. This photo from a stereo card shows a small mine and cut timber to shore up the tunnels.

Four old time miners sit in front of their cabin, located in the rugged Colorado hills. One plays a tune on the fiddle while another reads from a book.

A lone miner sits inside his log cabin. Note the ever present large jug of booze next to him on the floor.

This fine photograph shows the Homestakes Mine operation in Colorado. The newly completed headframe and buildings were ready for operation, as this image proves.

Piles of silver bars stand outside the warehouse of a Colorado mining operation. This stereo card by Russell did not identify the people in the shot.

This stereo photo shows early construction of a mine being dug out of the earth by an entire family. At the site, near Cripple Creek, Colorado, Grandpa is ready to lay a blow with the sledge hammer while the others offer encouragement.

This image by Standley, who photographed all over the town of Cripple Creek, was taken from high on a mountainside in 1900. It shows the vast area built up as a result of mining in the area.

A nicely composed photo stereo card, showing several men shoveling in rock to be crushed. In the background are several mines around Cripple Creek, Colorado.

William Henry Jackson (WHJ) was hired to glamorize much of Colorado for the tourist trade. Part of his series of photographs are displayed here. This view shows the Abe Lincoln Mine, in the Cripple Creek area. Piles of debris litter the hills.

These miners are working at the 1,000 foot level underground in a Cripple Creek mine. WHJ

The Mary McKinney Mine in the Cripple Creek District. WHJ

The Midget Mine in the Cripple Creek District. WHJ

A steam hoist used to lower and raise the miners' cage. This was photographed in the Cripple Creek District by Standley.

A busy street scene in a built-up mining town. On the far right in the background we can make out the mining buildings.

The Wild Horse Mine in the Cripple Creek District. WHJ

The town of Anaconda in the Cripple Creek District. WHJ

A view of Goldfield in the Cripple Creek District. WHJ

A distant view of Independence in the Cripple Creek District. WHJ

A bird's eye view of Victor in the Cripple Creek District. WHJ

A view of Leadville and the mining district. WHJ

The Gold King Mine in 1893. Photographed in Cripple Creek, Colorado by William Henry Jackson.

A mining property in the Cripple Creek District. The following three snapshots were taken by a private party who recorded the slow end of the Cripple Creek mining area in the 1920s.

Two small miners' houses in Cripple Creek.

By 1925, even large structures such as this aerial tram with empty railroad cars had been left behind by their owners. The tram is indicated by the arrow at the upper left.

Compare this scene with the earlier one of Goldfield taken by William Henry Jackson (page 105), and you will see that people not only have left, they have taken everything they could haul away with them. This is the last of the snapshots of the Cripple Creek Area.

The main street of a newly developing town in Colorado called Gillett, shown in 1895. The mining town has its welcome sign out for anyone to stay. Photo is by Harlan.

A group of miners in Colorado stop to pose for one of their own, who took this snapshot.

This photograph of the Bobtail Lode Mine, near Central City, Colorado Territory, might be the early work of W.H. Jackson.

At one time, Colorado's Tabor Matchless Mine made millions. Here it is shown "in ruins."

Thanks to the forethought of photographer W.G. Chamberlin, who worked out of Denver Colorado, we have much information about this stereo photograph. Standing in front of a cabin from the Dolly Varden silver mine are: (No.1) William Harding (cook); (No.2) Alvak Weller; (No.3) Wm Murphy; (No.4) Cotton (Montizuma); (No.5) Jas. Sullivan (foreman); (No.6) John Williams. The Dolly Varden mine was at the 18,500 foot level on Mount Lincoln. With the exception of the Montezuma, this was the highest habitation known. (This was written on the back of the stereo.)

At a mine in Colorado, photographer Standley has taken a shot of the steady stream of horse-drawn wagons that pick up or deliver supplies to the mine. Notice the steel rimmed wheels that hold up well carrying heavy loads.

A small snapshot of Denver, Colorado in the early days.

This photograph of the Mining Exchange building in Denver, taken by W.H. Jackson, is from the original print.

A rare photo of Idaho Springs, another mining town in Colorado. The photo is by WHJ.

A view of Colorado City taken by WHJ.

A photo postcard showing the small town of St. Elmo, Colorado. The sender of this card writes: "Dear Stella, it's 5:15 P.M. and all is safe and sound. We entered camp for the night, a mile above here at Saint Elmo. We killed a young cotton tail on our way up. Be sure and feed and water my bird. Well, everybody is ready to go. Fred. Aug. 6, 1919."

This photo of the stamp mill and gold concentrator at Ouray, Colorado, is from an original stereo card.

Chapter Four

Arizona

In early 1885, Arizona was making its mark in mining, as prospectors covered the state, finding good ore all over its range. Wickenburg, located at the fork of the stage road from Ehrenberg to Prescott, was one mining village of over 300 people. The town of Globe was found to contain mineral deposits of gold and silver, and gold was also discovered in the hills above a place called Tombstone. All of Arizona, in fact, has some type of ore—copper, gold, silver, iron, lead, tin, and platinum.

Some of the more important mining camps were The Humbug, The Tiger, Walnut Grove, Turkey Creek, Big Bug, and Lynx Creek, and these are in only one district. The Oro Blanco District is best known for the Ostrich Mine, which was a large gold mine only eighty miles from Tucson.

Tombstone, located twenty-seven miles from Benson, Arizona, was the county seat of Cochise County at that time. It was named after a miner who was joked into its name, telling him the only thing you'll find there is your tombstone. Because of all the mining in the area, Tombstone sported fully stocked stores and merchandise of all sorts, There were three churches, a schoolhouse, and two daily newspapers. The managers of the largest mines in Tombstone during this period were the Scheiffelin brothers, the Corbin brothers of Connecticut, and former Arizona governor Anson Safford.

The road from Tombstone to Bisbee passed through Charleston, a modest town on the San Pedro River. The Tombstone Mill had two 20-stamp mills at this point, one run by steam power and the other run by water power. Charleston had four stores, two hotels, several saloons, and a population of about 400 people, who did a large trade with Sonora. Smuggling was running rampant at this time, with Mexicans knowing all the little trails and roads of the areas. Mescal, tobacco, and cigars were brought from Sonora, while dry goods were taken across to Arizona.

Between Charleston and Tombstone is the old "Bronkow Mine," whose dark and bloody history has given it a permanent dark spot in southern Arizona. Discovered in 1885 by Emil Bronkow, a German engineer, its history contains an uninterrupted series of violence and murders. Bronkow and three companions were the first to be murdered there by Mexicans. Since then, more than seventeen men have been murdered on this property. Their graves lie thick around the adobe houses. Prospectors and miners avoided this spot like the plague. The gloomy old buildings were all unoccupied, and the new owners of the mine put up a house some distance away. It was, as mines go, not paying very much, and those that were there guarded the place with Henry rifles.

The first Arizona copper discoveries were made in 1883, and attracted little attention until the original miners sold off their claims to larger companies. By 1885, the little strike began producing large quantities of ore, and those investors turned a little scrap of paper into the Copper Queen mine. It soon had taken over an entire mountain.

This photo postcard offers us a look at a place called the Bar Room in Charleston, part of the Arizona Territory.

Calling himself "Arizona Bill," this prospector has his mules packed and ready to go.

This very rare photograph taken by A. Neumann shows the Yankee Mine in Arizona, owned by Phelps Dodge Company. In the foreground, Chinese labor workers are guiding pack mules up the hill. Seen below are small mine cars being loaded up from the large mine structure, and the huge head frame of the mine itself.

At Bisbee, Arizona, the Irish Mag Mines are shown operating in the hills and mountains of the area.

The Bird Cage Theatre is among the buildings shown in Burton Frashers' photograph of Main Street in the famous town of Tombstone.

The Tombstone Drug Store seems to be open for business in this later day photo card, also by Frashers. Across the street is the famous Crystal Palace saloon.

The Magma Mine, located at Superior, Arizona, looked like this to a photographer of the day.

From a photo postcard, this view looks down on the mining town of Jerome, Arizona.

Two views of miners working underground in the Warren Mining Districts of Arizona.

A photograph taken by one of the miners of the mine called Copperhill, in the hills near Jerome, Arizona.

The site of the Bagdad Copper Camp, Arizona. The mine is at the lower portion of the photo and the person who took the shot stood on top of the mine's headframe.

A photo postcard of the mine at Clarkdale, Arizona, sent on December 25, 1915.

A great photograph of the Jerome, Arizona, steam engine that serviced the town and mine. The miners' homes dot the landscape on the sides of the hills.

Jerome's large mining operation required the use of chemists to constantly check the ore for its contents. Here a metallurgists begins that task.

An ice plant in Parker, Arizona. The delivery wagon waits outside to begin loading.

The Phelps Dodge Smelter works near Douglas, Arizona are shown in this photo postcard.

Long winding train tracks lead to the Mammoth Mine site in Arizona. Electric power was used to run the mine cars out of the large holes in the mountain.

The main road leading to the town and mines of Morenci, Arizona.

A "bird's eye view" of Morenci. The miners' little homes dot the landscape.

This massive mine in Morenci was called the No 6. One wonders what the other five sites looked like.

The famous Tom Reed Mine and mill, near Oatman, Arizona.

Another view of the large Tom Reed Mine works.

This photograph is titled "Street Scene, Oatman, Arizona." It looks like a large meeting of owners and miners who may be losing their money.

A view from the hills overlooking the town of Oatman.

Situated in northern Arizona, the mining town of Hackberry had a train depot and a main street.

Chapter Five

New Mexico and Utah

By 1885, Silver City, New Mexico had a population of close to 6,000 people. This region was found to contain large quantities of gold bearing ore as well as silver ore. During this early period, the way to get there was via overland stage route from Lordsburg. The most promising area was the Black Range and good placer mining was begun there. Unlike mines in the northern states—which had to contend with harsh winters—those in New Mexico could be worked year around, so many of the miners flocked to this area, with Robinson City, New Mexico, the principal town in the range. Interestingly, there is no such name as the Black Range on any map of New Mexico, but what is there is the Mimbres Mountains. The ores found in this area are gold, copper, and silver. There was plenty of water in the area and abundant forest for timber. However, the warlike Apache Indians had occupied these mountains first, and held the key to expansion.

The first prospecting done in the Black Range was by Harry Pie of Georgetown, New Mexico, along with a friend. They followed the headwaters of the Cuchilla Negra and Rio Polomas and made a claim at several locations. The next year, Pie came back to the location with a Mexican helper, but both were killed by Apache Indians. It is safe to say that prospecting along these hills was dangerous until 1884, when the Indians were dislodged enough for miners to venture forward. By 1885, the miners had entrenched themselves with their numbers and the Apaches stopped their attacks.

In one area of Utah's Beaver County, which is 250 miles southwest of Salt Lake City, a large smelter was built in the same place where miners worked. The Newhouse Mines and Smelters owned two thousand acres of land that surrounded the rich Cactus Mine. The town itself had the Newhouse Opera House, a bank, a clubhouse, a rooming house, eating places and cafés, and eighty cottages with four to six rooms in them. In 1901, Mr. Samuel Newhouse, with some great daring, carried out the redemption of the Cactus Mine. He also purchased the rights to the Wah Wah Springs, in the mountains to the west, eight miles from the town. This was the only available water supply in that section of the country.

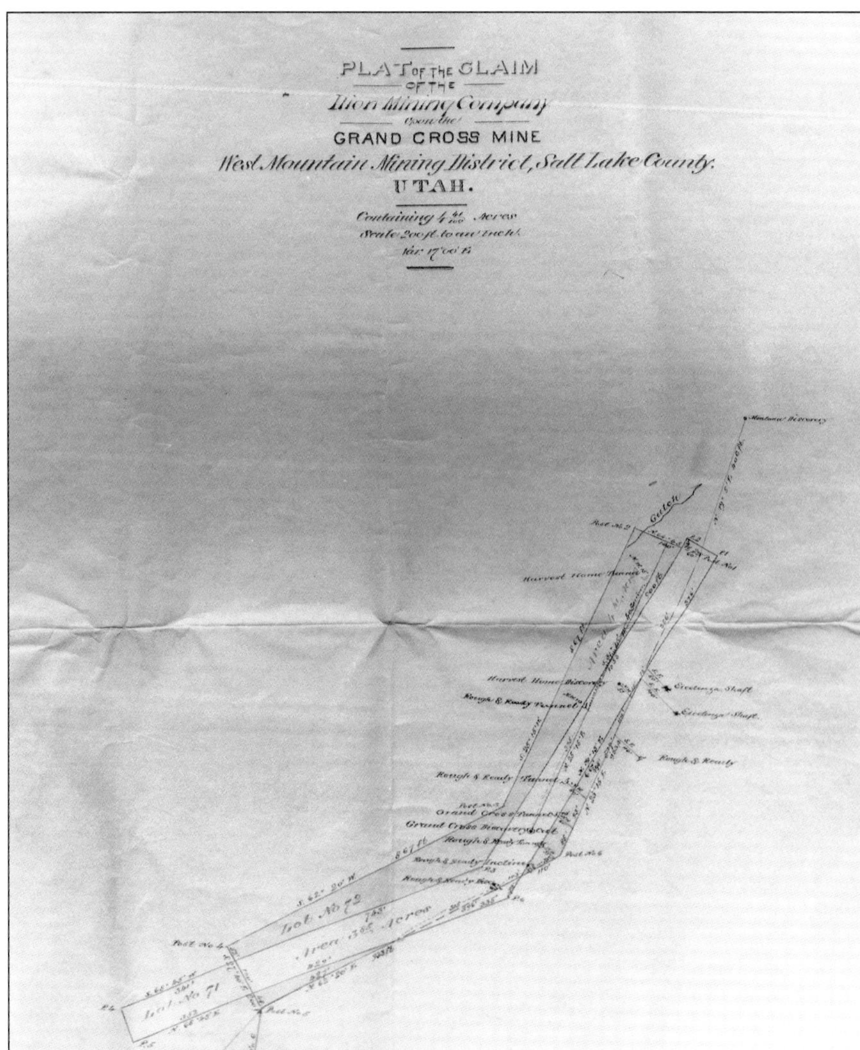

A photograph of the "Plat of the Claim" for the Grand Cross Mine, noted as being in the West Mountain Mining District of Salt Lake County. The drawing shows the boundaries and areas of the mining claim. Listed are the Rough And Ready, Grand Cross Discovery Cut, Harvest Home Discovery, and others.

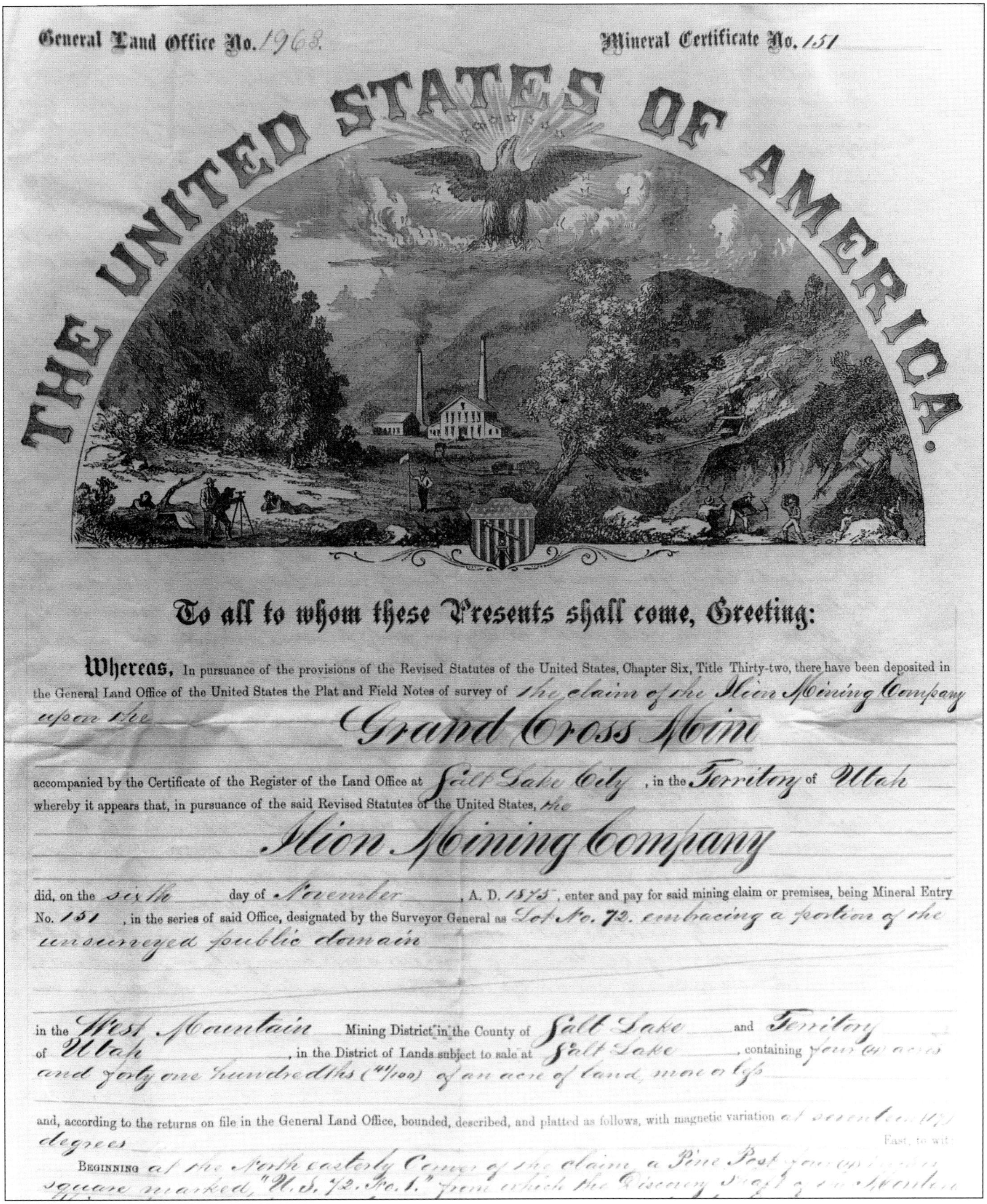

General Land Office No. 1968.   Mineral Certificate No. 151

THE UNITED STATES OF AMERICA.

To all to whom these Presents shall come, Greeting:

Whereas, In pursuance of the provisions of the Revised Statutes of the United States, Chapter Six, Title Thirty-two, there have been deposited in the General Land Office of the United States the Plat and Field Notes of survey of the claim of the Ilion Mining Company upon the

Grand Cross Mine

accompanied by the Certificate of the Register of the Land Office at Salt Lake City, in the Territory of Utah whereby it appears that, in pursuance of the said Revised Statutes of the United States, the

Ilion Mining Company

did, on the sixth day of November, A. D. 1875, enter and pay for said mining claim or premises, being Mineral Entry No. 151, in the series of said Office, designated by the Surveyor General as Lot No. 72. embracing a portion of the unsurveyed public domain

in the West Mountain Mining District in the County of Salt Lake and Territory of Utah, in the District of Lands subject to sale at Salt Lake, containing four (4) acres and forty one hundredths (41/100) of an acre of land, more or less

and, according to the returns on file in the General Land Office, bounded, described, and platted as follows, with magnetic variation at seventeen (17) degrees East, to wit:

Beginning at the North easterly Corner of the claim a Pine Post four (4) inches square marked, "U. S. 72. No. 1." from which the Discovery Shaft ...

This mining grant for the Grand Cross Mine in the territory of Utah dates from 1876. The cover of the three page certificate bears this fine engraved illustration of miners and engineers working on the landscape. The third page is signed by the President of the United States, Ulysses S. Grant.

This photograph of the Granite Mine in Utah shows the shaft of the mine in the foreground.

An early prospector panning for silver or gold in a stream.

Engineers pause for a moment to have their photograph taken. One is fully armed in this untamed area.

Extracting power from rushing water using wheels such as this was one of the engineer's jobs.

Out in the boondocks of Utah, these early miners have been working hard on the their small claim.

In Roswell, New Mexico, wood was as scarce as pretty girls. In this photo, a team of mules hauls in a wagonload of timber for heating houses.

In this photograph of Silver City, New Mexico, the large building in the foreground is the E. Cosgrove warehouse. Much of the town had been torn down due to a large flood in the area.

Called the Red Star Mine, this small mining operation in New Mexico may have paid off for the few miners who worked it.

It looks like the end of the line for this long water flume, as the mine in the background may have come up empty.

The houses in the background are on what was called Chihuahua Hill, in Silver City, New Mexico. The railroad steel bridge off the spur leads into town and the depot.

Wishing all a "Merry Xmas" from the Zinc Mine in New Mexico.

Miners eating lunch stop for a moment to have their portrait taken.

A fine example of an early general store selling everything from cigars to ice cream sodas.

Taking a break from a hard day's work, these fellows are playing a friendly game of cards.

Chapter Six

Oregon and Washington

During 1885, prospectors were roaming all over the west's territories, and some found their way to a place called Salmon River, in the Washington Territory. A town site was established, consisting of a hotel, saloons, shanty houses, and a sawmill. Between the two forks of the Salmon River and rising to over 4,000 feet is Mineral Hill. Supposedly there are quantities of silver and gold ore on top of this hill. It was there that Tacoma Concentration Company purchased two mines, The Homestead and the Toughnut Mines. Soon the area was flooded with prospectors going to Rock Creek and Skagit Mountain. The routes of travel were by stagecoach from Spokane Falls.

Gold quartz was also found up and down the Cascade Mountains and it was well known that the Tieton basin held a large quantity of this metal. Small discoveries were constant during the 1800s.

The following eight photographs are from a mining site called "The Blue Hog Mine" in Oregon. The photos and their descriptions were placed in a album. The description for this one reads: Upper tunnel "Blue Hog Mine" Magalla Ridge Con Mines Co. (Hill site in foreground).

Construction of the boarding and bunkhouse for the Blue Hog Mine.

Tunnel #2 at the Blue Hog Mine, with bunkhouse seen at the left.

Tunnels #1 and #3, from the platform of flume #1 thirty feet below.

Sleeping quarters for the Blue Hog Mine, employees out in front.

Reservoir of the Oregon Light and Power Company, located below the mining area.

The Blue Hog Mine's surveyor with his tools.

The logging mill where wood was ordered for the mine's construction

Site of the Red Boy Mine, located near Baker, Oregon.

A nice photograph of the Smigmi Mill and Mines, Gravillle Grant Company, Oregon.

Hidden against a hillside was this small mine. At the top of the photo we see the tailings of the mining operation.

The power station interior at a fairly small mining operation.

Hidden in a gully, the small mine shown here is only part of the overall mining operation.
operation.

The stamp mills of the small mine.

Exterior photograph of a small mining operation. A miner is pushing a cart of raw ore out of the hole in the mountain.

Portrait of a miner's family.

Three miners stand in front of this large mining operation, which has two tracks for mine cars and a foreman's house.

The whole town of Rockaway, Oregon has come out for a holiday celebration. The lone horseman on the right has to pick up the ring while at full gallop.

A nice photograph of a mine site, with the mill and headframe in the background. The long track leads to the mine itself.

Although famous for his logging and train photographs, Kinsey took this fine picture in 1900 of a mining operation deep in the Oregon woods.

Made from a glass negative, this photograph is a fine example of hydraulic mining using a heavy monitor.

An early photograph of Everett, Washington, called the "Pittsburgh of the West."

A splendid photograph of a early placer miner with his sluice box and long tom, showing how it's done.

Photograph of a unknown town in Washington. The thin paper print was just pasted on a board.

A nice photo of four miners who took the time to pose.

Positioned by their blockhouse, these army men with rifles stand guard over mining operations and any trouble that may occur.

A photograph showing water pressure at work. Notice the train engine's headlight (used for night work) in the right hand foreground.

A coastal town called South Bend, Washington in early 1900.

Up the river or over land by rail, people traveled west to hunt for riches. The paddlewheeler shown in this photo is making some real time.

Chapter Seven

North Dakota, South Dakota, Wyoming, Montana, Idaho

The Black Hills of North Dakota comprise an area that originally developed out of several treaties with the Indians, one of which was supposed to be the last with the Lakota or Sioux Indians in 1868. In the fall of 1874, a report by General Custer's expedition, which contained a mention of gold in the Black Hills, created a wave of excitement throughout the country, resulting in the formation of the Gordon Party at Sioux City, Iowa. Arriving at the Hills in the fall of 1876, they established camp at French Creek. They built an elaborate stockade to ward off Indian attacks, and several cabins to live in. Several of the party went prospecting for gold, which was found in sufficient quantity to justify further pursuit.

Their return to Sioux City and tales of gold in the Black Hills fueled the excitement. But this was opposed by the Indians, who didn't want whites on their sacred lands. Soon there were over 1200 trespassers on the Indian lands, and these had to be removed by government troops that General Crook had in his charge. Crook left behind seven men who would look after the interests in several mining camps. These men took to staying at the Gordon Stockade, where they remained until Captain Pollack escorted them to the Red Cloud Agency.

The Black Hills were not opened to settlement until February 28, 1877, and, as noted, all those that came in earlier were trespassers on the Indian reservations. There were gold miners in places like Bear Gulch, Custer, and Hill City, and when gold was discovered at Deadwood the mad rush to the placers began.

There was constant discord with the Indians, who would hide and pick off miners one by one if they could. Guns were worn on every man's belt, and long knives carried in case of fights in close quarters. Nights were especially dangerous, as the Indians would steal up to the camps, drive off the animals, and then massacre members of the party. Indians would attack small settlements, but would not try their hand on larger towns.

The reports about rich gold strikes in the hills near Deadwood Gulch brought a new influx of people, many leaving Custer City for the new Eldorado. At Lead City, which was a rather large town because of the many placers and the gold run, there was no trouble with the Indians. Lead City is about three miles from Deadwood and the home of the famous Homestake Mining Company. Brothers Mose and Fred Manuel, along with Jack Harney, discovered rich quartz there early in 1876 and named the area The Homestake. They also found claims that they named The Old Abe and the Terra. The Manuels continued developing the Homestake and Old Abe, and built a road to Kirk.

The first mill built on the Homestake by its new owners, Haggin, Hearst, and Tevis, was an 80-stamp mill, and it cost $30,000 to ship it from Cheyenne by ox teams to the Homestake Mine near Deadwood.

Early placer camps in the Black Hills had names like Hill City, Sand Creek, Potato, Mallory, Bear Gulch, and other Gulches all on the western edge of the state and into Wyoming. At Spring Creek, a flume was built that was fourteen miles long to carry water to the "dry diggings."

At this early time, all supplies had to be hauled great distances from the railroads, which were miles away. Mail was hauled by stagecoach and ox teams brought in supplies of food, water, and clothing for the harsh winters ahead. Highwaymen often attacked the stagecoaches and looted the passengers of their belongings. By 1890, an extension to Deadwood was finally built by the Elkhorn and Missouri railroad.

Other minerals found in the Black Hills included tungsten, mined near Yellow Creek; marble and onyx of beautiful tints found at various points in the outer hills; and cinnabar, found at Poisoned Ox mine, west of Rapid City. In addition, lead, gypsum, nickel, cobalt, and platinum were found in these hills.

Other Black Hills mining operations during and up to the year 1903 were The Imperial Gold Mining Company, The Golden Empire Mining Company, The Victoria Gold Mining and Milling Company, The Hidden Spanish Mining Company, and The Ruby Gold Mining Company, to name just a few.

There were many mines that brought riches in copper and silver in the state of Montana. Among them were the Boulder and Cooke City Mines with many scattered placer miners working the ravines and river beds along Boulder Creek. The Grub Stake Mine had a fine location for a mill and other mining operations.

In the 1880s, prospectors were also looking at a place called Castle Mountain because a new mine had opened there. This mining venture was said to produce more carbonate ores than even Leadville, Colorado. In addition, the elevation was not as extreme as in Leadville, where heights of 10,000 to 13,000 feet were common. Good wagon roads to haul timber and supplies to the camps were also available.

Since the opening of the Northern Montana Indian reservations in 1885, a great many miners had also gone to the Bear Paw Mountain range. There they found rich gold and silver veins to promote, and used hydraulics and ditches in their mining.

The area called Butte was a district from Meaderville to Rocker Montana in which the very ground was covered with quartz and veins popped out of the ground. The area was owned by hundreds of people who filed claims in the area. In 1888, millions of tons of quality ore were taken from this region.

Tourists riding the mine cars of a copper mine company in Idaho.

Hydraulic mining at Elevator Gulch in Wardner, Idaho. Photographed by T.N. Barnard.

These three photos are of the same mine site in the middle of winter. Here, some of the buildings are still under construction. The mine tailings can already be seen in the photograph, however.

The owners of the mine ride their two-horse cart during an inspection.

The mine in the snow, still under construction.

This marvelous photograph by W. J. Carpenter of Spokane, Washington, is of the Bunker Hill Mine in Wardner, Idaho. The view, taken from the winding dirt road, leads the viewer's eye through the town of wooden buildings and into the mining area in the distance.

Construction of the ore bin at the Orogrande Gold Mine, Clearwater, Idaho, in 1917.

A very well built but unnamed mine in Idaho is shown in this photo postcard.

A pack of mules hauling ore from a local mine.

A coal mine photographed near Williston, North Dakota.

The main street of Ambrose, North Dakota looked like this in 1905. The First National Bank, a pool hall, and a mercantile company are among the buildings seen.

This gold panner from Rockerville, South Dakota is telling his pals that they have it rich. The photo by Grabill is from 1889.

From a stereo card, this photo shows us the town of Custer, South Dakota, in 1909. The image is by well-known photographer Butcher.

Smoke from the blasting at the west portal of the Deep Creek tunnel mine site.

Far down in the tunnel of Deep Creek Mine, miners stand for their "flash powder" photograph in 1912.

A foreman and his aide stand on the nearly completed waterway for their mine site, way off in the background.

Checking the waterway by boat for any leaks.

L. Huffman, an early photographer from Montana, took this fine photo of a western cabin complete with potbelly stove, pots and pans, rifles, guns, and a chair. Construction of this cabin must have been done by several men as the roof beams are made of very large and heavy timbers.

A sign indicating the place where gold was first discovered in Montana is shown in this historical photo. The place was called Gold Creek, and the creek is seen to the right.

A wonderfully sharp photograph of a new town called Landusky, Montana. The mine's chemical tank can just be seen at the bottom of the photograph. Two X's mark the spot where the father had his stores and the O was where they lived.

This photograph shows Wallace Street in the gold mining town of Virginia City, Montana. The stagecoach has made a stop and becomes part of the picture.

A great photo postcard image of a short-lived town called Taft in the state of Montana. The railroad line was serviced by the Northern Pacific and it looks like folks are headed out of town. The postmark is from 1908 and the card bears this message: "Dear Rose, this is a picture of the famous Taft Town." The sender goes on about a tunnel camp, indicating there was mining in the area. Seen in the photo are theaters, saloons, and restaurants.

A good look at the cramped quarters miners had to endure while earning a living. They also had to live with the danger that at any moment that roof of rock above their heads could fall onto them.

A gambling bunch of miners who don't seem worried that they're sitting on dynamite boxes.

THE GWIN MINE HOIST.

A photo of the Gwin Mine tells us that this mine is going to go deep. Piles of large timbers for shoring up the ceiling and walls are waiting outside.

Interior shot of the Gwin Mill stamp mill areas.

A fine photo of Benton, in the Wyoming Territory. Seen are the T.A. Kent Wines and Liquors plus several other local stores.

Jack Morrow at Benton. Morrow seated on barrel surrounded by thirteen other men.

The man named Jack Morrow, seated on the barrel, must have been a rich man in Benton as he is the only one identified in the caption to this picture.

Unloading heavy mining equipment from a barge on the river. Waiting alongside the hoist are some railroad cars. If the lone small mining car on the lower left could talk, it might be saying something like "Don't drop that on me!"

It is remarkable that pioneer miners maintained the stamina to visit hundreds of sites all over the west looking for the "really big one"—a site that would lead them to the easy life they so desired. Sadly, it was the large companies who held the cards. They could afford the big stakes needed to dig up the earth and find that pot of gold. The lone miners who, along with a friend or two, dug up rocks to uncover the mother lode did not generally reap the profits that were just over the next hill.

To conclude this book on *Mining in the Old West*, I leave this fine photograph of Oregon miners and their horse, who stopped to have their portrait taken by a unknown photographer.

Appendix

Department of the Interior, U.S. Geological Survey Map, 1910

Included here is an eight-section U.S. Geological Survey Map showing portions of southern California and Nevada (including the Death Valley area where summer temperatures reach over 120 degrees). The full map is shown here, and the individual map sections are shown in greater detail on the following two pages. Sections 1 to 4 represent the top portion, left to right; Sections 5-8 represent the bottom portion, left to right.

Shown on the map are some of the desert towns and mine sites illustrated in this book. For example, Skidoo, the California mining town that lasted only a few years, can be seen in the upper right corner of section 6 on page 172. (A rare newspaper from this town is shown in Chapter One, on page 38). Section 3 on page 173 shows the Bullfrog District of Nevada, where much of that state's mining took place. Additional points of interest on the map are as follows:

Section 1, page 172: Gerro Gordo Mine
Section 3, page 173: Goldbar, Bullfrog, Rhyolite, Gold Center
Section 4, page 173: Calico Hills, Horn Silver Mine
Section 5, page 172: Soda Works Springs, near Keeler Ophir Mine
Section 6, page 172: World Beater Mine, Burro Mine, Modoc Mine, Skidoo Mine
Section 7, page 173: Gold Hill, Borax Mill

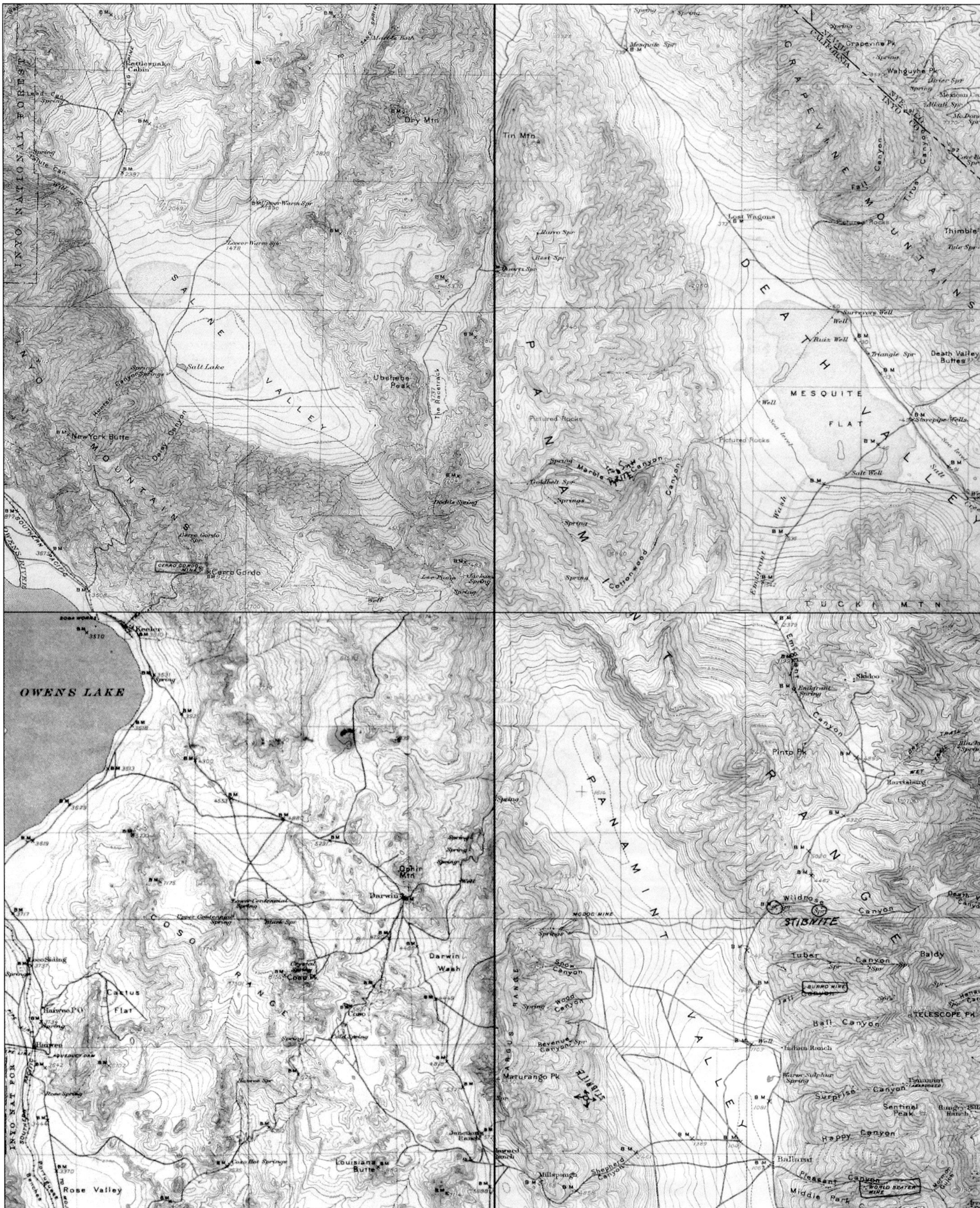
INYO NATIONAL FOREST
Rattlesnake Cabin
Dry Mtn
Upper Warm Spr
Lower Warm Spr
SALINE VALLEY
Salt Lake
Ubehebe Peak
The Racetrack
INYO MOUNTAINS
New York Butte
Daisy Canyon
Dodds Spring
Cerro Gordo
Keeler
OWENS LAKE
Darwin
Ophir Mtn
COSO RANGE
Darwin Wash
Coso Pk
Cactus Flat
Rose Valley
Louisiana Butte
Tin Mtn
Mesquite Spr
GRAPEVINE MOUNTAINS
Grapevine Pk
Wahguyhe Pk
Lost Wagons
Fall Canyon
Titus Canyon
Thimble Pk
DEATH VALLEY
Surveyors Well
Ruiz Well
Triangle Spr
Death Valley Buttes
MESQUITE FLAT
Stovepipe Wells
Salt Well
Pictured Rocks
Marble Canyon
Cottonwood Canyon
PANAMINT
TUCKI MTN
Emigrant Canyon
Emigrant Spring
Skidoo
Pinto Pk
Wildrose Canyon
STIBNITE
MODOC MINE
Tuber Canyon
Baldy
BURRO MINE
Jail Canyon
Hall Canyon
TELESCOPE PK
PANAMINT VALLEY
ARGUS RANGE
Snow Canyon
Wood Canyon
Revenue Canyon
Maturango Pk
Surprise Canyon
Sentinel Peak
Happy Canyon
Ballarat
Pleasant Canyon
WORLD BEATER MINE
Middle Park Canyon
Shepherd Canyon

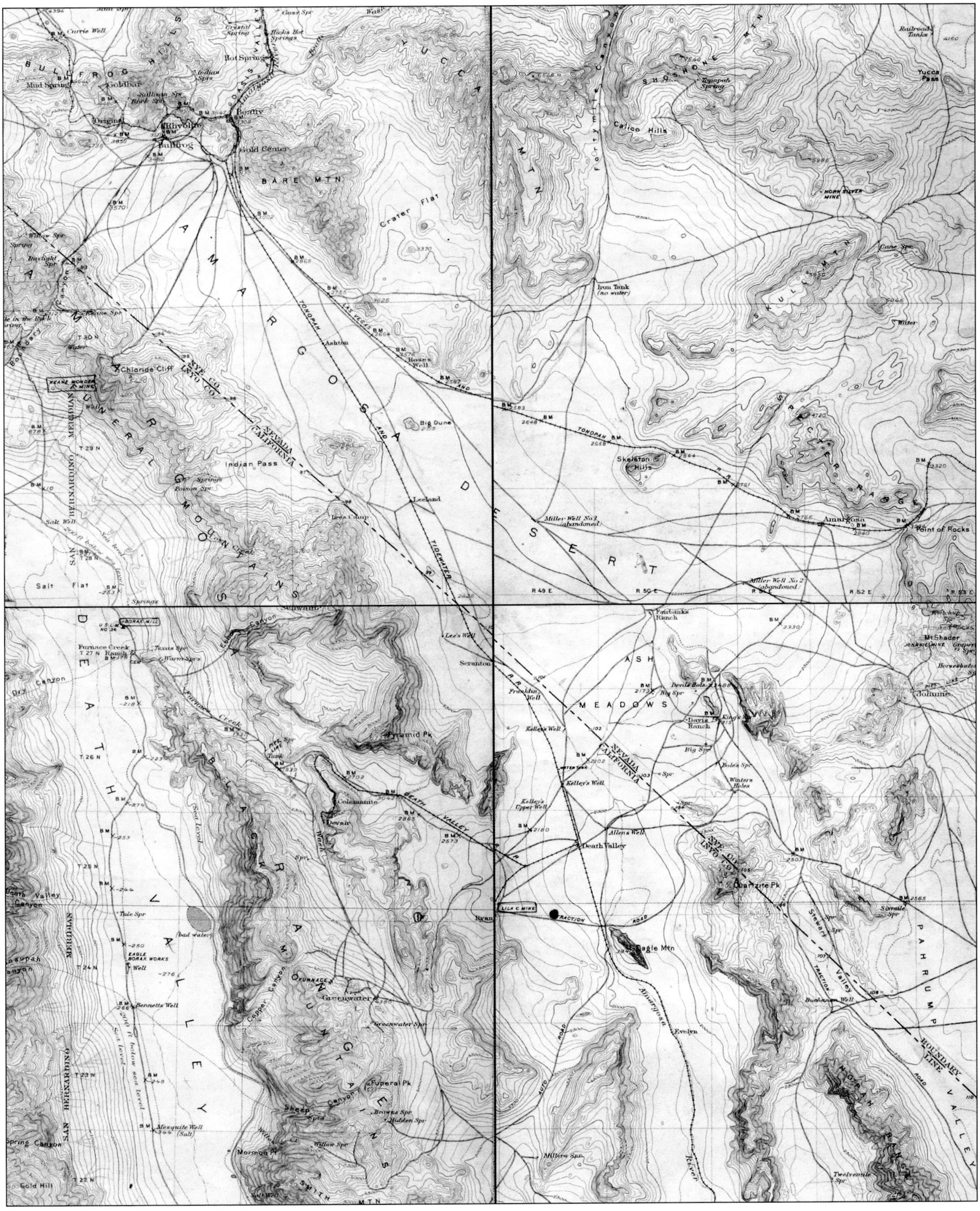

BULLFROG HILLS
Currie Well
Mud Spring
Goldbar
Original
Rhyolite
Bullfrog
Beatty
Gold Center
Hot Spring
Crystal Spring
Hicks Hot Springs
BARE MTN
Crater Flat
YUCCA
AMARGOSA DESERT
TONOPAH
LAS VEGAS
TIDEWATER
Ashton
Big Dune
Indian Pass
Leeland
Chloride Cliff
KEANE WONDER MINE
FUNERAL MOUNTAINS
NYE CO.
INYO CO.
NEVADA
CALIFORNIA
SAN BERNARDINO MERIDIAN
T 30 N
T 29 N
T 28 N
T 27 N
T 26 N
T 25 N
T 24 N
T 23 N
T 22 N
Salt Well
Salt Flat
Springs
SHOSHONE MTN
Topopah Spring
Calico Hills
Forty Mile Canyon
Railroad Tanks
Yucca Pass
HORN SILVER MINE
SKULL MTN
Cane Spr
Iron Tank (no water)
SPECTER RANGE
Skeleton Hills
Miller Well No 1 (abandoned)
Miller Well No 2 (abandoned)
Amargosa
Point of Rocks
R 49 E
R 50 E
R 52 E
BORAX MILL
Furnace Creek Ranch
Texas Spr
Furnace Creek
Dry Canyon
DEATH VALLEY
Pyramid Pk
Colemanite
Ryan
Lee's Well
Scranton
Franklin Well
ASH MEADOWS
Fairbanks Ranch
Devils Hole
Davis Ranch
King's Spr
Big Spr
Bole's Spr
Winters Holes
Kelley's Well
Kelley's Upper Well
Allens Well
Death Valley
Johnnie
Mt Shader
Quartzite Pk
LILA C MINE
TRACTION ROAD
Eagle Mtn
Amargosa River
Evelyn
AUTO ROAD
Stewart Valley
Buckman Well
Sixmile Spr
PAHRUMP VALLEY
BOUNDARY LINE
NOPAH RANGE
Twelvemile Spr
Millers Spr
BLACK MOUNTAINS
AMARGOSA RANGE
Tule Spr
bad water
EAGLE BORAX WORKS
Bennetts Well
Copper Canyon
Greenwater
Greenwater Spr
Funeral Pk
Sheep Canyon
Browns Spr
Hidden Spr
Willow Spr
Mormon Pk
Mesquite Well (Salt)
Spring Canyon
Gold Hill
SMITH MTN

Glossary of Mining Terms

Drifting. The act of following along with the course of the vein.
Adit. Open cut or tunnel on the vein.
Amalgam. Gold or silver combined with mercury.
Apex. The top of the highest point of a vein.
Auriferous. Any rock or sand bearing gold.
Bedrock. Rock underlying placer mines.
Blind lode. Where there appears no outcrop to the vein.
Breast. The face of a tunnel or drift.
Breccia. Angular rocks cemented together.
Cage. An elevator used in hoisting ore.
Cap Rock. Rock underlying the ore or vein.
Cobbin. Breaking ore for sorting.
Contact Vein. A vein between two dissimilar rock masses.
Country Rock. The rock on each side of the vein.
Crosscut. A level driven across the course of a vein.
Dip. The slope or pitch of a vein.
Drift. Horizontal passage underground, driven on or along the vein.
Face. The end of a drift or tunnel.
Fault. Displacement of a stratum or vein.
Feeder. A small vein entering into a larger one.
Float. Loose ore or rock separated from the original formation.
Fissure Vein. A crack in the earth containing a mineral.
Footwall. Layer of rock immediately beneath the vein.
Free Milling. Ores that separate from the gangue by simple methods.
Gash Vein. A vein wide at the top and closing at a short depth.
Hanging Wall. The layer of a rock or wall overhanging a lode.
Headframe. The structure that surmounts a mineshaft. It supports the pulley for the mine's elevator, which moves up and down the shaft
Horse. A mass of rock between the branches of a vein.
In Place. Not having been distributed from its original position.
Level. A horizontal passage in a mine, diverging from the shaft.
Lode. A vein, thus a lode claim is a class having in its boundaries, a vein. The term is sometimes used to designate the mine itself.
Mill Run. A test of the vein of a given quantity of ore.
Open Cut. An open surface working.
Outcrop. The portion of a vein showing at the surface.
Placer Mining. Early panning for gold flakes hidden among the rocks and dirt of streams.
Pyrites. Brittle cubes or iron sulphide.
Stamp Mill. A machine that uses heavy weights to crush ore.
Strike. The extension of a lode in a horizontal direction.
Stulls. A framework to support rubbish when stopping out ore.
Seams and stringers. Used interchangeably for narrow branches running off from the vein or lead.
Tailings. Huge mounds of debris, rocks, and dirt left behind after ore was crushed and sifted.

Bibliography and Credits

Baldwin, George P. *The Black Hills Illustrated*. Black Hills Mining Association, 1904.

Calaveras County California, Mining Resources, 1848-1898.

Christians, William F. Articles dated 1932.

Crofutt's New Overland Tourist Pacific Coast Guide of 1880.

Geologic Guide Book for the Mother Lode Country. State of California Dept. of Natural Resources, Division of Mines, San Francisco, California, September 1948.

Lee, Willis T., Ralph W. Stone, Hoyt S. Gale, and others. *Guide Book of the Western United States, Part B. The Overland Route*. U.S. Geological Survey Bulletin 614, Government Printing Office, 1916.

McGinnis, Jack. *Randsburg Mining Area*. 1984.

Report of the Secretary of War on the Geology and Typography of California, May 17th 1850.

Scheeline, Jacqueline Snyder. *Historical Sketch of Genoa [Nevada]*. 1949.

Shankland, S. R. *Early Black Hills History*.

The Cripple Creek District, Colorado, Geology & Mining Industries, U.S. Geo. Survey.

The Cripple Creek Story, 1910.

The Overland Guide 1885. Rand McNally.

Weight, Harold and Lucile. *Rhyolite, City of Golden Dreams*. The Calico Press 1966.

The West Shore Publication of 1888, & News about the Pacific North West.

Credits

Photographs of Bodie and Aurora, California panoramics by Forbes and Frashers, thanks to Pete Korngiebel, Bishop, California.

Photographs of mine and graveyard in Virginia City, Nevada, thanks to the Nevada State Historical Society.

All others from the Demlinger collections.

Would you like to write a book?

For over thirty years we have had the privilege and pleasure of meeting hundreds of people with wonderful ideas for new books. For many of them, writing a book was only a far off dream, something they hoped to do someday. With our encouragement they went on to produce a beautiful, useful addition to the literature in their field.

If you are one of those people who always dreamed of writing a book, we would love to hear from you. New ideas and authors are the lifeblood of what we do here. We have gathered together a team of people who will do everything they can to make the concept of a book become a reality.

We are eagerly seeking authors to write illustrated, regional history books. With vintage photographs and postcards these big, 11" x 8-1/2" books are bringing bygone eras back to life. Illustrated in glorious full color, each book is stunning & unique.

The following form has helped hundreds to gather their thoughts and begin the process of compiling a book. Submit your idea to us in one of the following ways:

• Email the following information to info@schifferbooks.com

• Copy and mail this form to Acquisitions, Schiffer Publishing, Ltd. 4880 Lower Valley Rd., Atglen, PA 19310

• Fax us at 610-593-2002

We look forward to hearing from you!

Name ______________________________

Street Address

____

____

____

Home Phone ______________________________

Work Phone ______________________________

Fax ______________________________

Email address: ______________________________

Website: ______________________________

Best time to call: ______________________________

The subject of the book you are proposing: ______________________________

____

____

____

____

Your background in the field: ______________________________

____

____

____

____

____

The number of postcards/historic images you own or have access to: ______________________________

A description of your target audience, i.e. population, relevant organizations, etc.: ______________________________

____

____

____

____

____

Specific retailers who would carry your book:

____

____

____

____

____

____

____

____

____

____

Media that would review/promote your book:

____

____

____

____

____

____

Schiffer Publishing, Ltd.
4880 Lower Valley Road
Atglen, PA 19310
610-593-1777
www.schifferbooks.com